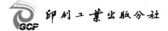

高职高专"十四五"印刷专业规划教材

平版制版双语培训特色教材

平版计算机直接制版技术

（中英文）

主　编｜陈　斌　项建龙

PINGBAN JISUANJI
ZHIJIE ZHIBAN JISHU

文化发展出版社
Cultural Development Press
·北京·

图书在版编目（CIP）数据

平版计算机直接制版技术：汉、英 / 陈斌，项建龙主编 . — 北京：文化发展出版社，2024.5
ISBN 978-7-5142-4336-9

Ⅰ．①平… Ⅱ．①陈… ②项… Ⅲ．①计算机辅助制版－汉、英 Ⅳ．① TP391.72

中国国家版本馆 CIP 数据核字（2024）第 080993 号

平版计算机直接制版技术（中英文）

主　　编：陈　斌　项建龙	
出 版 人：宋　娜	
责任编辑：李　毅　杨　琪　雷大艳	责任校对：岳智勇
责任印制：邓辉明	封面设计：韦思卓

出版发行：文化发展出版社（北京市翠微路 2 号 邮编：100036）
发行电话：010-88275993　010-88275710
网　　址：www.wenhuafazhan.com
经　　销：全国新华书店
印　　刷：北京九天鸿程印刷有限责任公司
开　　本：787mm×1092mm 1/16
字　　数：450 千字
印　　张：21.75
版　　次：2024 年 6 月第 1 版
印　　次：2024 年 6 月第 1 次印刷
定　　价：68.00 元
ＩＳＢＮ：978-7-5142-4336-9

◆ 如有印装质量问题，请与我社印制部联系。电话：010-88275720

序

中华文明源远流长，五千年历史从未中断；泱泱大国代有创造，数不尽文化辉煌硕果。印刷术作为中华优秀传统文化中的一颗璀璨繁星，为人类文明的远程传承和赓续发展起到了巨大的推动作用。随着时代的更迭和科技的发展，人类正在由印刷文明走向数字文明，中国印刷迈进了高质量发展的新时代，正朝着"绿色化、数字化、智能化、融合化"转型升级。近年来，中国印刷技术和工艺上的创新层出不穷，为中国印刷的持续发展提供了强大动力。

印刷技术升级以印刷制版技术的革新为基础。随着人工智能技术的兴起，印刷制版技术正向智能化快速发展，并将成为未来十年中国印刷技术智能化发展的重点之一。中国印刷制版企业瞄准技术变革和产业优化升级，以科技创新为动力，以产品质量求生存，时时刻刻都在进步，创造出计算机直接控制的智能制版系统，并凭借优质的创新能力、稳定的产品特质、贴心周到的售后服务，为全球印刷业的发展提供新的动力。当前，中国印刷制版设备的技术水平和服务能力，已经达到国际先进水平，在国际市场占有一席之地。

为适应印刷技术数智发展的趋势，满足全球印刷业对高素质、高水平、高层次印刷制版技术技能人才的需求，更好地服务印刷技能人才培养，助推智能化印刷制版技术的发展与普及，上海出版印刷高等专科学校联合杭州科雷机电工业有限公司，发挥各自优势，共同编写了《平版计算机直接制版技术》教材。本教材供行业教学培训使用，是一个服务教学前沿的好事。本教材内容包含平版印刷印版的种类、制版技术工艺、计算机直接制版技术，印版显影的后处理技术等，涵盖了平版印刷制版的各个技术工艺控制环节。教材编写以平版印刷制版职业岗位能力培养为主线，参考了"印前处理和制作员"与"印刷设备维修工（计算机直接制版机CTP）"的中国国家职业技能标准，紧扣平版印刷制版工作岗位的主要操作内容，着力塑造专业基础技术操作能力和综合技术应用能力，适用于具有一定印刷专业基础知识，从事平版印刷和制版工作的专业人士使用。

  本教材的编写由中国印刷职业教育机构与创新优质的中国印刷企业联合完成。教材整合领先技术，内容丰富，科学实用，以技术、知识、专业精神引领行业发展，有助于推动印刷业数字化、智能化升级，特别是该教材采用双语版，更有助于推动全球印刷业的人员沟通、技术流通和数字贯通。在此，我对教材编辑出版者表示感谢，衷心希望世界各国的印刷企业和专业人士以结识本教材为契机，积极参与印刷人才教育培养领域的交流，建立长期密切合作机制，共同助力全球印刷业创造更加美好的未来，使千年不衰的印刷业为人类新文明再立新功。

柳斌杰

2024.5

前 言

平版制版是将原稿制作成平版印刷印版的工艺过程。平版印刷的原理是基于油水不相溶的特性：空白部分具有良好的亲水性能，吸水后能排斥油墨，印刷部分具有亲油性能，能排斥水而吸附油墨；先将印版用水润湿，使空白部分吸附水分，再上油墨；空白部分吸附水而不能吸附油墨，印刷部分则可吸附油墨，并通过转移油墨形成图文信息。常用的平版有PS版（预涂感光版）、平凹版、蛋白版（平凸版）、多层金属版等，每种印版的表面均由亲油疏水的图文部分和亲水疏油的空白部分组成。平版制版工艺从最初的以石板制版为基础的技术，经过不断发展，技术已经非常成熟。

平版计算机直接制版是一种数字化平版印版成像技术，出现于20世纪90年代，1995年德国德鲁巴印刷技术及设备展览会（Drupa）首次展出了42种平版计算机直接制版系统。平版计算机直接制版技术采用数字化工作流程，直接将文字、图像转变为数字，用计算机直接控制，通过激光扫描成像，显影处理后生成直接可上机印刷的印版。平版计算机直接制版技术省去了人工拼版、晒版等工序过程。采用计算机和数字技术的平版计算机直接制版具有速度快、效率高、工序少、流程简化、可变因素少、全程数字技术处理、稳定性提高、管理规范等优点。

本教材是"印前处理和制作员"的中国国家职业技能标准的重要组成部分，全书共有九章，以平版印刷制版岗位能力为主线展开，分别从平版制版的基础知识、平版计算机直接制版工艺过程与制版设备操作要点、打样与标准化管理几个方面对平版计算机直接制版工艺与技术进行了全面的说明。本教材强调知识内容为操作技能服务的宗旨，实践性强，适用具有一定印刷专业基础知识的从事平版制版工作的专业人员，同时也可以作为平版制版专业技术指导，以及平版制版职业岗位考核学习参考用书。希望阅读本书后，读者能对平版计算机直接制版技术有一个全面理解，并结合实际工艺流程中的应用提高对平版制版技术的全面掌握。

　　本教材为双语教材，出版之际恰逢国家主席习近平提出"一带一路"倡议十一周年。十一年来，"共建'一带一路'成为深受欢迎的国际公共产品和国际合作平台"，习近平主席在党的二十大报告中强调要继续"推动共建'一带一路'高质量发展"。"一带一路"倡议的坚持共商协作、实现共同增长的原则，将指导中国印刷产业在积极推进"一带一路"国际合作中发挥应有作用。

　　在此还要感谢在本教材编写中给予了极大帮助与支持的上海出版印刷高等专科学校的王丹、田全慧等老师，同时杭州科雷机电有限公司的陈炜、范燮军、李纯弟、Victor Wong、姚文、董海、张丽仙等也积极地参与了本教材的编撰与出版工作。

　　开卷有益，希望本教材能给读者提供一些实际的帮助。同时，欢迎广大读者对书稿编写的不足之处提出宝贵意见和建议，以便本教材修订时补充更正！

编者

2024 年春

目 录

第一章 绪 论

第一节 平版制版分类与流程 / 1
第二节 传统 PS 版制版原理与流程 / 2
第三节 平版计算机直接制版原理与流程 / 3

第二章 平版制版基础知识

第一节 直接制版机的基本工作原理 / 14
第二节 输出分辨率及其设置 / 15
第三节 加网的作用及其设置 / 17
第四节 平版制版输出文件格式 / 19
第五节 字体技术 / 23
第六节 拼版的基本概念与方法 / 26
第七节 CTP 加网技术 / 28
第八节 印刷补偿曲线和反补偿曲线基本知识 / 31

第三章　印版输出准备工作

第一节　生产作业的环境要求 / 35
第二节　输出流程软件与设备驱动软件的安装 / 37
第三节　印版输出的线性化 / 40
第四节　输出流程软件操作 / 43
第五节　制版机及外接设备 / 62
第六节　冲版机 / 67

第四章　印版的输出

第一节　CTP 印版输出 / 70
第二节　设置印版的分辨率与加网参数 / 78
第三节　印版输出过程中出现的问题的解决方法 / 83

第五章　印版的显影

第一节　CTP 印版的显影 / 88
第二节　CTP 印版的显影问题解决 / 90

第六章　印版质量的检测

第一节　CTP 常见质量问题 / 98
第二节　印版测量仪器 / 99
第三节　印版输出检测 / 100
第四节　CTP 印版成像质量控制 / 102
第五节　网点增大的原理及控制 / 106

第七章　制版设备测试与异常处理

第一节　CTP 印版输出与测试 / 111
第二节　异常情况分析及处理 / 115
第三节　判定印版输出质量的方法 / 120
第四节　印前处理及制版与印刷质量的关系 / 126

第八章　检查数字打样样张

第一节　样张质量检测与控制方法 / 127
第二节　数码打样测控条 / 129
第三节　测量仪器的使用方法 / 135

第九章　平版印刷标准与管理

第一节　相关印刷质量标准 / 139
第二节　平版制版设备配置标准 / 143
第三节　工艺流程控制 / 145
第四节　生产与环境管理 / 149

参考文献

第一章 绪 论

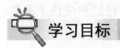 学习目标

1. 掌握平版制版的基本概念与方法；
2. 了解传统 PS 版制版的基本工艺流程与特点；
3. 了解平版计算机直接制版的基本工艺流程与特点；
4. 掌握平版计算机直接制版工作的基本流程。

第一节 平版制版分类与流程

平版制版是将原稿制作成平版印刷印版的工艺过程。平版印刷印版上的印刷部分和空白部分几乎在同一平面上，空白部分具有良好的亲水性能，吸水后能排斥油墨，印刷部分具有亲油性能，能排斥水而吸附油墨。印刷时，先用水润湿印版，使空白部分吸附水分，再上油墨，因空白部分已吸附水分，不能再吸附油墨，而印刷部分则吸附油墨，印版上印刷部分有油墨后便可印刷。常用的平版有预涂感光板（PS 版）、平凹版、蛋白版（平凸版）、多层金属版等。每种印版的表面均由亲油疏水的图文部分和亲水疏油的空白部分组成。

根据使用的材料和方法不同，有许多种平版制版方法。

一、分类

1. 按版材分类

平版制版的版材有石板、锌板、铝板、纸基版等类型。

石板是最早的平版印刷版材，石板版材笨重，只能直接印刷，所以现在已不再使用。

锌版和铝版是现今常用的版材。纸基版是近年来发展起来的版材，常用于静电制版，数量不太多的印刷品印刷以及轻印刷。

2. 按制版方式分类

平版制版按制版方式分为手描版、转印版、即涂版、预涂版、多层金属版、平凹版、静电版、干式版等。其中手描版是原始的平版制版方法，是用汽车墨手工描绘在版材上形成印刷部分，现在已不再使用。

3. 按印版表面结构分类

印版按表面结构分为平版、平凸版、平凹版。

二、平版制版工艺流程

平版制版的关键技术是将印版的表面处理为亲水疏油的空白部分与亲油疏水的图文部分。平版制版工艺从最初的以石版制版为基础的工艺，经过不断发展，已经形成技术非常成熟的工艺，主要有传统的预涂感光版（PS版）和计算机直接制版两种工艺。

第二节　传统 PS 版制版原理与流程

一、传统 PS 版

PS 版是预涂感光版（Pre-Sensitized Plate）的缩写。

PS 版的版基是 0.5mm、0.3mm、0.15mm 等厚度的铝板。铝板经过电解粗化、阳极氧化、封孔等处理，再在版面上涂布感光层，制成预涂版。

二、PS 版分类与流程

PS 版按照感光层的感光原理和制版工艺，分为阳图型 PS 版和阴图型 PS 版。

阳图型 PS 版的制版工艺过程如下：

曝光→显影→除脏→修版→烤版→涂显影墨→上胶

曝光是将阳图底片有乳剂层的一面与 PS 版的感光层贴合在一起，放置在专用的晒版机内，真空抽气后，打开晒版机的光源，对印版进行曝光，非图文部分的感光层在光的照射下发生光分解反应。常用的晒版光源是碘镓灯。

显影是用稀碱溶液对曝光后的 PS 版进行显影处理，使见光发生光分解反应生成的化合物溶解，版面上便留下了未见光的感光层，形成亲油的图文部分。显影一般在专用的显影机中进行。

除脏是利用除脏液，把版面上多余的规矩线、胶粘纸、阳图底片粘贴边缘留下的痕迹、尘埃污物等清除干净。

修版是将经过显影后的 PS 版，因种种原因需要补加图文或对版面进行修补。常用的修补方法有两种：一种方法是在版面上再次涂上感光液，补晒需要补加的图文；另一种方法是利用修补液补加图文。

烤版是将经过曝光、显影、除脏、修补后的印版，表面涂布保护液，放入烤版机中，在 230～250℃的恒定温度下烘烤 5～8 分钟，取出印版，待自然冷却后，用显影液再次显影，清除版面残存的保护液，用热风吹干。烤版处理后的 PS 版，耐印力可以提高到 15 万印以上。如果印刷的数量在 10 万印以下，不必对 PS 版进行烤版处理。

涂显影墨是将显影墨涂布在印版的图文上，这样可以增加图文对油墨的吸附性，同时也便于检查晒版质量。

上胶是 PS 版制版的最后一道工序，即在印版表面涂布一层阿拉伯胶，使非图文的空白部分的亲水性更加稳定，并保护版面免被脏污。

PS 版的砂目细密，图像分辨率高，形成的网点光洁完整，具有良好的阶调、色彩再现性。

清除使用过的 PS 版版面上残存的油墨和感光层，在原来的铝版基上重新涂布感光液，形成新的感光层，便可重新制成打样版或正式印刷版。这种利用用过的 PS 版的铝版基重新制作 PS 版的方法叫作 PS 版的再生，PS 版的铝版基可重复使用，因此 PS 版是平版印刷中使用最多的印版。

第三节 平版计算机直接制版原理与流程

一、平版计算机直接制版

平版计算机直接制版技术出现于 20 世纪 90 年代，1995 年德鲁巴印刷技术及设备展览会（Drupa）展出了 42 种计算机直接制版（CTP）系统。在 2000 年德鲁巴印刷技术及设备展览会（Drupa）上，来自世界各地的 90 多家计算机直接制版系统及材料生产商展出了近百种产品。

计算机直接制版是一种数字化印版成像技术。计算机直接制版是采用数字化工作流程，直接将文字、图像转变为数字，用计算机直接控制，用激光扫描成像，再通过显影生成直接可上机印刷的印版，省去了胶片这一材料、人工拼版的过程、半自动或全自动晒版工序。CTP 的特点是采用计算机直接制版工艺（无胶片）、速度快、效率高、工序少、流程简化、可变因素降低、全程数字技术处理、信息量增加、稳定性提高、管理规范。

计算机直接制版采用数字化工作流程，直接将文字、图像等作业转变为数字信号，用计算机直接控制制版设备扫描成像，再通过冲版机显影生成印刷的印版。图 1-1 为 CTP 工艺流程。

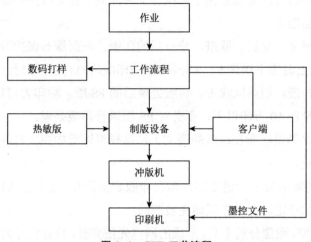

图 1-1　CTP 工艺流程

制版机的结构原理如图 1-2 所示，采用 16/32/48/64/96/ 或是 16～96 个之间任由配置且独立的 830/NM/1W 或 405/NM 半导体激光作为光源，通过光纤耦合把 N 个光源导到密排面上，密排面上射出的激光通过光学镜头聚焦并紧密吸附在光鼓表面的版材上，对版材进行热烧蚀或感光。工作时装有版材的光鼓做高速旋转运动，而装有光纤密排系统及光学镜头系统的扫描平台做横向同步运动。驱动电路系统根据计算机的点阵图像来驱动各独立的激光器进行高频开关，从而在版上形成点阵图像潜影。

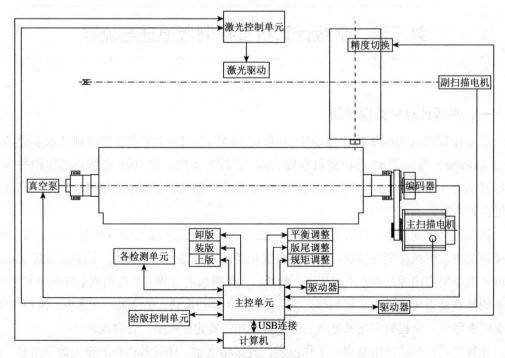

图 1-2　制版机结构原理

二、平版计算机直接制版工作流程

1. 平版制版的工作流程

平版制版的工作流程总体经过五个环节。

（1）印版输出前的准备。通过输出流程软件进行印版输出参数与模版设置。

（2）印版输出。通过制版机的操作实现平版印版的激光曝光与输出。

（3）印版的显影。曝光后的印版通过显影机，将印版表面处理为可以上机使用的平版印版。

（4）印版质量检测。针对制版输出的印版的图文进行观察，并检测相关的区域与标识，判断印版的制版质量。

（5）印版的后处理。为了提高印版的印刷适性，经曝光显影后的印版还需要经过烤版、上胶等后处理工序。

2. 印版输出前准备（以佳盟优联软件为例）

（1）流程软件的工作步骤

①启动佳盟优联服务器

佳盟优联是建立在 C/S 工作模式上的。服务器是整个流程的核心，客户端与服务器进行交互操作。客户端的每一步操作都经过服务器来进行处理，服务器控制客户端的运行。输出器与服务器进行传输操作，文件经服务器解释并传输到输出器，然后在输出器中进行相关输出操作。

启动佳盟优联服务器是运行佳盟优联流程的第一步操作，只有当服务器成功启动后，相应的功能模块才能开启，才能执行相应的各工序处理，客户端程序也才能运行登录。

注意：在启动服务器前，首先要保证网络连接正确，与之相连的计算机，特别是客户端计算机应该可以正常访问这台服务器。

在桌面上，双击"Server"图标或在"开始"菜单中选择"JoinUs Unity"下的"Server"程序，即可启动佳盟优联服务器。如果安装的是限时试用版，流程服务器启动时，将首先弹出提示流程加密锁使用期限的对话框，上面会明确提示整个系统的使用期限。

继续单击"确定"，即可打开服务器界面。

在界面上，可以查看服务器已登录用户的信息及功能模块是否成功启动。服务器界面分为左右两部分。

客户端连接信息，左侧显示登录到服务器的客户端信息，包括 ID、用户名、IP 地址等。

ID：登录序号，按登录顺序自动生成；

用户名：登录的用户名称或输出器的名称；

IP 地址：登录的客户端或输出器所在主机的 IP 地址；

工单处理器信息：右侧显示服务器已启动的功能模块；

连接数据库成功：表示数据库已成功启动并与服务器正常连接，如果显示"数据库连接失败"则流程无法正常工作；

规范化模块已启动：表示规范化处理器模块已成功启动；

拼版模块已启动：表示拼版处理器模块已成功启动；

输出模块已启动：表示输出处理器模块已成功启动。

如果运行时没有检测到加密狗或者软件使用的授权码与当前使用的加密狗不对应，将提示"加密狗错误"。

②启动佳盟优联客户端

客户端是用户的操作端口，用户在客户端提交规范化、拼版、输出等操作到服务器，并将每一步处理后的结果显示在客户端中。

只有在流程服务器成功启动后，用户才能启动客户端。

用户在登录客户端时，首先要确定这台计算机和服务器所在计算机连接正常，客户端与服务器端最好设置在同一个局域网内，每台计算机都拥有自己的IP地址，这样客户端才能更好地访问服务器。其次，客户端在一台计算机上只能启动一个，即不允许多个客户端同时在一台计算机上操作；当用户在不同的计算机上启动客户端时，不能重复使用同一个用户名登录，即同一个用户同时只能登录一次。

在桌面上，双击"Client"图标或在"开始"菜单中选择"JoinUs Unity"下的"Client"。稍后即可打开客户端登录对话框，如图1-3所示。

图1-3　客户端登录对话框

在对话框中，输入正确的用户名、密码、服务器IP地址，单击"确定"按钮，即可打开佳盟优联客户端。如果输入的IP地址不对，将提示"连接服务失败"，单击"确定"按钮，退出登录界面。佳盟优联客户端成功登录后，在"优联服务器"界面上，显示出该用户登录的信息。

③启动佳盟优联输出器

佳盟优联输出器是流程中的重要组成部分，负责整个流程的输出工作，接收服务器解析的页面文件，并控制文件向相应的设备进行输出。

由于流程中同一类型的输出器按照IP地址管理，因此在每一台计算机上只能启动一个输出器，并且输出器可以与服务器在同一台计算机上启动。

佳盟优联输出器分为CTP输出器和数码印刷输出器。双击桌面上相应的佳盟优联输出器图标或单击"开始"→"程序"→"JoinUs Unity"→"CTP输出器或数码印刷输出器"，打开输出器启动界面。

稍后即可打开相应的输出器登录对话框。在对话框中输入正确的服务器IP地址，所设 IP 地址为服务器所在计算机的 IP 地址，这样才能成功地与服务器连接。单击"确定"按钮，即可打开相应的输出器界面。

启动前，要确定两台计算机可以正常访问。

输出器成功登录后，在"流程服务器"界面上，将显示出该输出器登录的信息，表示相应的输出器已成功登录到服务器端。

（2）流程软件的工作过程

①建立作业

首先需要新建一个作业，为每个作业输入一个唯一的工单号。创建作业完成后，在流程服务器的文件系统中就会产生相应的作业文件夹，如图1-4所示。

图 1-4　新建作业

根据工作需要，给该作业添加若干个处理模板，通常至少需要一个规范化模板，还要对模板进行相关设置。这些包含 JTP 参数的处理模板就构成了该作业的"数字工单"。

工单号：工单号是整个系统中作业的唯一标识，不能与其他作业的工单号重复；

作业名：该作业的名称；

处理模板：为作业选择一个在模板库中定义好的流程模板；

客户信息：该作业所属的客户信息和资料工单号是必填项。作业名可与工单号名称相同，也可以为其命名其他名称。它们的名字中都不能包含下列字符：/ ? \ | * : " " < >。

②设置作业处理模板

"处理模板"界面的主要功能是建立当前作业所需的处理模板。用户可根据本次作业的需要来建立所需模板。新建作业时如果预设了处理模板，将显示在处理模板界面中，如图 1-5 所示。

模板是流程各个处理节点的参数设置数据，每一个节点就是流程中一个具体工序。作业处理模板中保存着处理这个作业所需要的各个节点模板。想要使用一个处理节点，就需要先把一个节点模板添加到作业中。

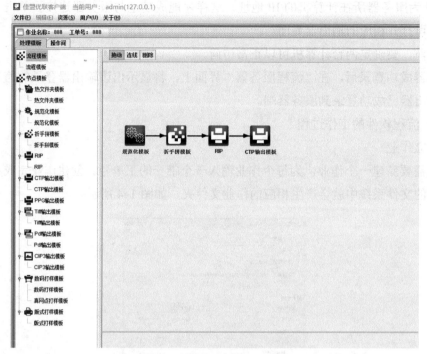

图 1-5　打开作业时的处理模板

　　a. 添加源文件

　　设置好处理模板后，用户要向该作业中添加源文件。流程会将添加的源文件上传到服务器的作业文件夹中，同时在工作时的源文件区域列出这些源文件。

　　b. 规范化文件

　　用户将源文件提交给规范化模板，触发流程服务器上的规范化处理器开始工作。该 JTP 根据规范化模板中的参数，读取相应的源文件并进行运算处理，生成规范化后的 PDF 文件。规范后的 PDF 文件以单页形式存在，称为"规范页"。规范页数据同样被保存在流程服务器上的作业文件夹中。

　　注意：从以上描述可以看出，在规范化过程中，客户端仅仅发出指令，所有执行都在流程服务器中完成，即由服务器的 CPU 执行规范化程序，从服务器中读取源文件并进行处理，将处理的结果存放回服务器中。整个过程中，没有任何页面数据在服务器和客户端工作站之间传递，这就是典型的客户端/服务器系统的工作模式。

　　c. 建立页面列表

　　经过规范化后的规范页可以在页面列表窗口中被整理成页面列表，可以由不同源文件的规范页共同组成一个页面列表。每个页面列表可以理解成一本有页码顺序的书。

　　③制作版式

　　整理好的页面列表，需要赋予它一个折手版式，以便进行折手拼版。折手版式是一个独立的数据文件，该文件记录了各个单独页面在大版上的排列位置和顺序。佳盟优联流程中内置了专门的版式设计程序，可以用它来预先设计好各种拼版版式。

a. 折手拼版

给页面列表设定好版式后，就可以提交给"折手拼版处理器"进行折手拼版操作。这个操作同样在流程服务器上得到执行。处理器读取各个小版单页数据，根据版式要求，把它们拼成一个大版文件，这个大版文件同样是一个 PDF 文件，也同样保存在流程服务器的相关作业文件夹中。

拼版完成的大版文件，会出现在"印张样式"窗口中。在这个窗口中可以进行整个版面的预览检查，检查完后就可以提交给输出器进行输出。

b. 输出印版

优联流程为每个输出设备配备了一个输出服务器，输出服务器是一个单独运行的程序，可以安装在不同的计算机上。想要使用一个输出服务器输出，需要在流程作业中添加一个此类设备的输出模板，并且将模板中的 IP 地址指向相应的输出服务器程序所在计算机。然后，向这个输出模板提交拼好版的大版文件，文件就会被输出到指定的输出服务器中。当输出页面出现在输出服务器的作业列表中后，说明该页面包括光栅化、加网在内的所有印前处理都已执行完毕。在输出服务器中还可以进行最后的精细预视检查，包括进行网点预视和叠印检查，检查完毕就可以放心地将其输出到印版上进行印刷，或输出到数码印刷机上打印出成品。

3. 印版输出

直接制版机的整个工作流程如图 1-6 所示。

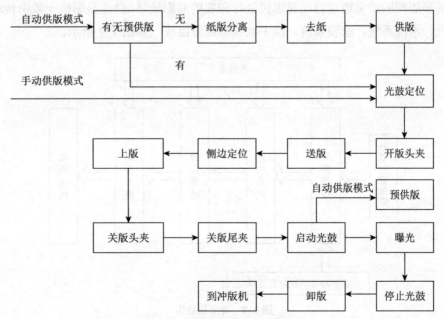

图 1-6 直接制版机的工作流程

模板建立：参数设置（版材幅面、精度、曝光功率）。

供版：如果是自动供版模式，首先判断是否已预供版，如果无预供版，则吸嘴下移，并判断是纸或版，是纸则去纸并返回执行前次动作，是版则供版到主机入版口；如果是手动模式或已供版，则跳过前面动作。

装版：先进行光鼓定位，光鼓定位在装版位置后，打开版头夹，随后送版系统把版送到版头夹处进行前定位，侧拦规系统对版材侧边进行定位，接着闭上版头夹，然后装版辊压下，光鼓转动装版，等转到版尾位置时，打开版尾夹，光鼓反转，把版尾送入版尾夹板下，关闭版尾夹，装版辊抬起。

成像：启动光鼓，此时若是自动供版模式，则同时启动供版机再次供版，等到光鼓转速达到设定值时，且达到真空压力后曝光开始，光学平台上的激光系统，将携带有计算机点阵信息的激光光束照射在版材上，实现文字和图像信息的传递。光学平台做匀速直线移动，滚筒旋转一周，平台移动对应光路数线距离，点阵信息结束，即完成一版曝光，而后，光鼓减速停止。

卸版：光鼓停止后再次定位，卸版机构开始进行卸版，排版机构将版材送入过桥，版经过过桥进入冲版机。

冲版：根据版材的显影温度、时间，进行显影、水洗、上胶、烘干等操作。

4. 印版的显影

曝光后的印版通过冲板机对印版进行显影处理。冲板过程中使用的显影药水具有化学腐蚀性和沉淀积聚性。

（1）设备结构

目前平版制版通常通过自动显影机进行印版的显影处理。自动显影机一般由传动系统、显影系统、水洗系统、涂胶系统、烘干系统等部分组成，如图1-7所示。

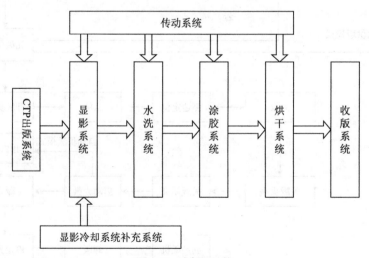

图1-7 冲版机结构

① 传动系统

传动系统是引导印版运行的驱动装置。电机通过涡轮蜗杆或链轮链条带动传送辊转动，并通过对压胶辊驱动使印版通过显影机的各个工作环节。

②显影系统

印版显影过程中，显影液槽通过加热器、冷却系统、循环过滤系统确保了显影液能在一个恒定的条件下工作，当印版通过显影槽时，毛刷辊在一定压力和速度下对印版进行刷洗，印版上的感光层快速溶解。

③水洗系统

水洗系统由两组喷淋管向印版正面和反面喷淋清水，以清洗掉显影生成物和残留的显影液，并通过挤压辊挤去版面的水分。该系统具备二次水洗的功能。

④涂胶系统

涂胶系统是清洗后的印版上涂布一层薄薄的保护胶，已确保在印刷前，印版的清洁和抗氧化。

⑤烘干系统

烘干系统由送风管和胶辊组成，通过吹风和加热使印版迅速干燥。

（2）调试操作说明［以虎丘影像（苏州）有限公司 PT 系列印刷冲版机为例］

安装完成后进入机器的调试阶段，首先向显影药槽内注入清水，直到清水从溢出口溢出为止（如果机器带有水洗内循环功能，水洗槽中也需加入清水至溢出为止），同时检查一下机器是否处于水平位置。显影补充液和胶水可先用清水代替。

需要注意的是：在注入清水的同时不能让水溅入电器部件内，如有请用电吹风吹干为止，否则绝对不能通电试验。

通电开机，首先利用操作界面设定必要的工艺参数，如显影温度、电机速度、版材尺寸规格等。在工艺参数设定完成以后，选择相应的程序让机器进入工作状态。机器进入加温状态，与此同时，仔细观察循环泵工作正常与否，具体可看液体的流动状态或用温度计测量显影温度是否上升。建议用户在正式调试以前对各项冲洗功能进行手动测试，并在正式运行后要观察各项功能是否正常。

冲版机各档轴的压力大小，直接影响出版质量，所以调好各档轴的压力至关重要，但由于使用的版材和药水差异很大，对各档轴的压力要求也不尽相同，所以请参考表 1-1 中的压力数值范围，根据实际使用版材和药水的要求，调节各档轴的压力至合适范围以满足实际出版需求。

测量方法：从版厚 0.27 mm 的版材上裁剪一条宽 30 mm 的版条，用合适的方法将拉力计的一端与版条一端固定，然后将版条的另一端从被测的一对轴中间插入，来测量辊轴转动时的拉力。

测试完成后，关闭冲版机主电源，将冲版机中的清水排放干净（过滤器中的水也需要排放干净），放完水后，注意要将调节后的阀门复位，然后向显影槽中加入显影液直至溢出为止，若有水洗内循环功能且要求使用此功能，请在水洗槽中加入清水直至溢出为止。将显影补充桶和胶水桶里的清水分别换为显影补充液和胶水。

表 1-1　冲版机参数

轴序号	拉力	备注
显影 1（橡胶轴）	60～70 N	测试环境： 温度 28℃ 湿度 50% 版材规格： 宽 30 mm， 厚 0.27 mm
显影 2（毛刷轴）	3.7～3.8 N	
显影 3（毛刷轴）	3.7～3.8 N	
显影 4（橡胶轴）	80～90 N	
水洗 1（橡胶轴）	60～70 N	
水洗 2（毛刷轴）	3.5 N	
水洗 3（橡胶轴）	80～90 N	
上胶 1（橡胶轴）	80～90 N	

注：表内的数据都是在轴表面干燥时测得的。

开机，正式冲版测试。根据客户实际使用的药水和版材的要求，设置好冲版工艺参数（显影温度、显影时间、毛刷转速等），待机器达到冲版要求后，正式冲版，根据出版的测试结果，对冲版机相关功能进行微调（如显影时间、辊轴压力、上胶量大小，等等），使版材冲洗效果达到要求。

（3）注意事项

仔细阅读产品的主要参数和技术指标，确认用户冲版要求是否与设备指标相符。

每天启动设备后，应注意观察设备工作状态是否正常。

水洗水流量应经常注意观察，尤其在水压不稳定地区，必要时安装流量压力表，每天结束工作时应将水排干净。

不要将宽度和长度不符合表格所列的版材规格以及卷曲的版材送入机器内进行冲洗，一旦发生版材在设定的时间内没有从出版口出来，应立即按急停按钮进行检查以免发生卡版现象。

按照提供的保养方法，必须定期进行清洗和保养。

每天结束冲版工作时，宜在待机状态下关机，依次关闭电源开关、空气开关，最后切断总电源闸，关闭水龙头。千万不可直接拉电源闸，这将有损机器的使用寿命。

严禁非专业人员私自开启电气箱，以免造成人身危险和设备故障。

5. 印版的后处理

显影处理后的印版，需要经过检查与测试。确认印版没有错误，并且质量符合标准后，需要对印版进行后处理。目前很多冲版机可以直接进行印版的后处理，即通过上胶与烘干系统实现。

（1）烤版

烤版是将经过曝光、显影、除脏、修补后的印版，表面涂布保护液，放入烤版机中，在230～250℃的恒定温度下烘烤5～8分钟,取出印版,待自然冷却后,用显影液再次显影,清除版面残存的保护液,用热风吹干。烤版处理后的印版,耐印力可以提高,但是如果印刷的数量不高就不必对印版进行烤版处理。

注意事项:

①因烤版之后的感光膜吸附在版面很牢固,所以PS版在烤版之间一定要将版面上的胶带、脏点清除干净;

②烤版保护胶擦得不宜过多,过多易出现留痕状痕迹,严重时不易着墨;

③修版后用清水将修版液冲洗干净,否则烤版后,残余的修版液及被溶解的物质会污染版面,引起上脏;

④烤版时必须待版面保护胶干燥之后才可进行烤版;

⑤涂烤版宇宙保护胶要用脱脂纱布,以免用脏布涂擦,污染版面,引起上脏;

⑥涂烤版保护胶液用力不要过大,以免纤维脱落而影响烤版质量。

(2)上胶

印版上胶可以保护版基表面细小的砂目,增强砂目的耐磨性,提高耐印力;可以保护版基的亲水层,提高图文部分的着墨性,有助于印刷时快速达到水墨平衡;可以保护印版,避免轻微刮花或划伤;可以有效封住砂目不让灰尘进去,防止印版直接暴露在空气中引起印版过快氧化。

上胶是平版制版的最后一道工序,即在印版表面涂布一层阿拉伯胶,使非图文的空白部分的亲水性更加稳定,并保护版面免被脏污。

第二章 平版制版基础知识

学习目标

1. 直接制版机的基本工作原理；
2. 输出分辨率及其设置；
3. 加网的作用；
4. 掌握平版制版文件格式及格式特点；
5. 理解平版制版后端字体；
6. 掌握页面拼大版的基础理论；
7. 理解CTP加网的网点技术；
8. 掌握印版补偿曲线的相关理论。

第一节 直接制版机的基本工作原理

一、直接制版机及其结构

计算机直接制版是一种数字化印版成像技术。计算机直接制版机由精确而复杂的光学系统、机械系统，以及电路系统三大系统构成。

光学系统：激光器、光纤耦合器、密排头、光学镜头、光能量测量等。

机械系统：机架、光鼓、墙板、送版部、版头开闭部、装版辊部、卸版部、排版部、丝杠导轨部、扫描平台部等。

电路系统：编码器、主副伺服电机、各执行机构步进电机、真空泵、主控板、接线板、激光驱动板、各位置传感器等。

由激光器产生的单束原始激光，经多路光学纤维或复杂的高速旋转光学裂束系统分裂成多束（通常是200～500束）极细的激光束，每束光分别经声光调制器按计算机中图像信息的亮暗等特征，对激光束的亮暗变化加以调制后，变成受控光束。再经聚焦后，几百束微激光直接射到印版表面进行刻版工作，通过扫描刻版在印版上形成图像的潜影。经显影后，计算机屏幕上的图像信息就还原在印版上供胶印机直接印刷。

二、直接制版机的种类及成像原理

计算机直接制版系统从曝光系统方面可分成内鼓式、外鼓式、平板式3大类；从版材品种方面可分为银盐版、热敏版（烧蚀式热敏版、非烧蚀式热敏版）、感光树脂版和聚酯版（非金属版基）等；从技术方面可分为热敏技术（普通激光成像）、紫激光技术、UV光源技术等。

1. 重氮型版材

重氮型版材通常用作平印版材。重氮版材中光化学分子数，直接与光照量和重叠的部分吸收有关，感光范围局限于光谱的紫外区域，无扩大效应，适用于大功率的氩离子激光器曝光。

2. 光聚物型版材

光聚物型版材的性能与重氮型类似，也无扩大效应，使用的光聚物的种类很多。机理与重氮型版材十分相同，用于制备平印版。曝光后形成交联型图像适用于紫外激光器曝光。

3. 银盐感光版材

银盐感光版材的机理是利用光化学反应，将银盐转化成金属银，以及这些银颗粒的催化扩大，促进周围银离子的还原。

4. 感光抗蚀剂版材

感光抗蚀剂版材与光聚物型和重氮型版材一样，感光范围通常局限于紫外部分，柯达生产的感光抗蚀剂MC929晒制平凹镁版用，可用氩离子激光和可见光激光器曝光，得到优质图像。

第二节 输出分辨率及其设置

1. 设备像素

印版在制作过程中通过激光曝光，形成印版上的线条、文字与网点，每一个曝光点就是一个"设备像素"。

2. 输出分辨率

印版在单位距离内输出设备像素的数量称为"输出分辨率"，以点/英寸（dots per

inch，dpi）为单位。

设备像素越密集，打印出来的图像越精细。换言之，输出分辨率越高，输出就越精细。2400 dpi 是现在输出系统默认的输出分辨率，可以满足精细印刷的需要。以 2400 dpi 输出的曲线非常光滑，而且小号字也很清晰。

3. 输出分辨率与加网线数的关系

加网线数（lpi）仅仅是加网部分的分辨率，输出分辨率（dpi）是整个印版的分辨率。

在一张印版上可以有不同的加网线数，如一个印版上整体使用 175 lpi，但是其中一张图片需要使用 100 lpi 做特殊效果，但是输出时整个印版使用同一个输出分辨率进行输出。加网线数比输出分辨率小得多。彩色印刷通常是 175 lpi、2400 dpi。

在一定范围内提高加网线数能使加网部分更精细，但对不加网的实地、文字和线条没有影响。提高输出分辨率既让实地、文字与线条精细，也让网点更饱满，从而提高加网部分的输出质量。因此，纯文字印刷品需要足够的输出分辨率而不是加网线数，图文并茂的印刷品两者都要。例如，纯文字的加网线数可以是 80 lpi，如果有插图，就要至少 133 lpi，但是输出分辨率都是 2400 dpi。输出公司针对各类业务将输出分辨率统一设定为 2400 dpi。加网线数与输出分辨率的关系为

$$加网线数（lpi）=\frac{输出分辨率（dpi）}{网目调单元宽度（点或者设备像素）} \qquad (2-1)$$

例如，输出分辨率为 2400 dpi，一个网目调单元的宽度为 16 点（即 16 个设备像素），则加网线数为 2400/16=150 lpi。

在输出分辨率固定的情况下，若要增加加网线数，则只能减少网目调单元可以容纳的设备像素数量。即：

$$网目调单元宽度（点）=\frac{输出分辨率（dpi）}{加网线数（lpi）} \qquad (2-2)$$

4. 输出分辨率与图像分辨率的关系

输出分辨率（dpi）与图像分辨率（ppi）的关系有以下三种情况需要考虑。

（1）对于全部加网的图像，只根据加网线数设置图像分辨率（通常是 300 ppi），设计师必考虑输出分辨率。

（2）对于全部不加网的图像，图像分辨率越接近输出分辨率，打印质量就越高。通常输出时就 2400 dpi，若将图像的分辨率设成 2400 dpi，将获得最佳的打印效果（只有在这种情况下输出分辨率和图像分辨率才等同，一个位图像素对应一个设备像素）。但是如果 2400 dpi 使图像文件过大，也可选择 900 dpi、1200 dpi 等。

（3）对于部分加网的图像，加网部分只要求 300 ppi，不加网部分却要求 900 ppi 以上。如果完全照顾不加网部分，可能使文件过大，这时可取一个中间值，比如 600 dpi。

第三节　加网的作用及其设置

一、网点

印刷采用网点再现原稿。印刷成品放大，就会发现它的图文是由无数个大小不等的网点组成的。网点越大，印刷出来的颜色越深；网点越小，印刷出来的颜色越浅。

网点的排列位置与大小是由加网线数决定的，例如，加网点数为150lpi，即1英寸的长度或宽度上有150个网点。不同颜色的网点会按不同的角度交错排列，以免所有颜色的油墨叠印在一起。

二、加网

在印前处理中，使用"加网"的方法，用一定数量的像素组合呈现明暗变化的"网点"，一个网点可以由不同大小阵列的像素构成。

加网方法目前主要有调幅加网与调频加网。目前，绝大多数平版印刷工艺都是使用调幅加网，调幅加网由三个加网参数控制。

1. 网点形状

印刷中的网点形状是以50%的着墨率情况下网点所表现出的几何形状来划分，可以分为方形网点、圆形网点和菱形网点三种。图2-1为显微镜下显示出的印刷网点形状。

方形网点在50%的覆盖率下，呈棋盘状。它的颗粒比较锐利，对于层次的表现能力很强，适合线条、图形和一些硬调图像的表现。

圆形网点无论是在亮调还是在中间调的情况下，网点之间都是独立的，只有暗调的情况下才有部分相连，对于阶调层次的表现能力不佳，四色印刷中比较少采用，如图2-1所示。

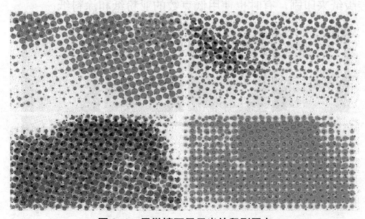

图2-1　显微镜下显示出的印刷网点

菱形网点综合了方形网点的硬调和圆形网点的柔调特性，色彩过渡自然，适合一般图像、照片。

2. 网点角度（如图 2-2 所示）

印刷制版中，网点角度的选择有着至关重要的作用。选择错误的网点角度，将会出现莫尔条纹。

常见的网点角度有 90°、15°、45°、75°等。45°的网点表现最佳，稳定而又不显得呆板；15°和 75°的角度稳定性要差一些，不过视觉效果也不呆板；90°是最稳定的，但是视觉效果太呆板，视觉美感较差。

两种或者两种以上的网点成一定角度叠印在一起，会产生一定的透光和遮光效果，当角度较小时会产生莫尔条纹，严重的会影响印刷图像的美观和质量，这种条纹俗称"龟纹"。

一般来说，两种网点的角度差在 30°和 60°的时候，整体的条纹还比较美观；其次为 45°的网点角度差；当两种网点的角度差为 15°和 75°的时候，产生的条纹会影响印刷图像质量。

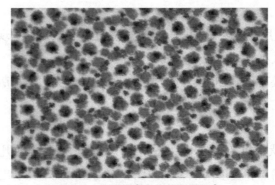

图 2-2　不同网点角度的圆形网点

3. 网点线数

网点线数的大小决定了图像的精细程度，如图 2-3 所示。常见的网点线数有以下几种。

10～120 线：低品质印刷，远距离观看的海报、招贴等面积比较大的印刷品，一般使用新闻纸、胶版纸来印刷，有时也使用低克数的亚粉纸和涂料纸。

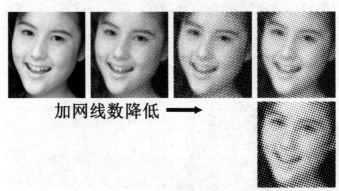

图 2-3　加网线数与图像精细程度

150 线：普通四色印刷一般都采用此精度，各类纸张都有。

175～200 线：精美画册、画报等，多数使用涂料纸印刷。

250～300 线：质量要求最高的画册等，多数用高级涂料纸和特种纸印刷。

第四节　平版制版输出文件格式

一、页面描述语言

页面描述语言从 20 世纪 80 年代诞生以来，得到了迅速发展和广泛应用。页面描述语言主要有 Adobe Systems 公司的 PostScript、Xerox 公司的 Interpress、Image 公司的 DDL 和 HP 公司的 PCL 5 等。其中最著名、应用最广的当数 PostScript 语言（以下简称 PS 语言）。

PS 语言于 1976 年诞生于美国的益世（Evans & Sutherland）计算机公司，当时是作为电子印刷的一种页面描述语言而设计的，后来几经修改，直到由查理斯·格什克（Charles Geschke）和约翰·沃诺克（John Warnock）创建于 1982 年的 Adobe Systems 公司再次实现这类语言时才被正式定名。该公司于 1985 年推出了第一台配有 PS 语言解释器的 Apple 激光印字机。经过短短几年，PS 语言就得到了广泛应用，并成为高质量图文印刷输出不可缺少的重要组成部分。1990 年，Adobe 公司又开发出了面向彩色文件的、功能更强的 PostScript Level 2 和面向工作站多窗口环境的 Display PostScript（即显示版 PS）。由于 PS 语言卓越的性能和广泛的应用，它已经成为电子出版行业事实上的工业标准。Adobe 公司通过扩展 PS 语言为不同层次的用户提供完整的打印解决方案，以此改进 PostScript 标准。新一代的页面描述语言有广泛的应用范围，从家庭和小型办公室到集团公司、从打印设备制造商到专业印刷行业都能适用。

二、页面描述文件与特点

EPS 是 Encapsulated PostScript 的缩写，是一种基于 PostScript 语言的矢量图形文件格式。EPS 格式是能在 Adobe Illustrator 和 Adobe Photoshop 软件之间相互交换的文件格式。EPS 文件是目前桌面印刷系统普遍使用的通用交换格式当中的一种综合格式。EPS 文件格式又被称为带有预视图像的 PS 格式，它是由一个 PostScript 语言的文本文件和一个（可选）低分辨率的由 PICT 或 TIFF 格式描述的像素组成。EPS 文件就是包括文件头信息的 PostScript 文件，利用文件头信息可使其他应用程序将此文件嵌入文档。

EPS 文件格式的"封装"单位是一个页面。另外页面大小可以随着所保存的页面上的物体的整体边界来决定，所以它既可用来保存组版软件中一个标准的页面大小，也可用来保存一个独立大小的对象的矩形区域。

其文本部分同样既可由 ASCII 字符写出（这样生成的文件较大，但可直接在普通编辑器中修改和检查），也可由二进制数字写出（生成的文件小，处理快，但不便修改和检查）。

EPS 文件虽然采用矢量描述的方法，但亦可容纳点阵图像，只是它并非将点阵图像转换为矢量描述，而是将所有像素数据整体以像素文件的描述方式保存。而对于针对像素图像的组版剪裁和输出控制参数，如轮廓曲线的参数、加网参数和网点形状、图像和色块的颜色设备特征文件（Profile）等，都用 PostScript 语言方式另行保存。

EPS 文件有多种形式，如按颜色空间有 CMYK EPS（含有对四色分色图像的 PostScript 描述部分和一个可选的低分辨率像素）、RGB EPS、L*a*b EPS。另外不同软件生成的各种 EPS 文件也有一定区别，如 Photoshop EPS、Generic EPS、AI（EPS 格式的 Illustrator 软件版本）等。在交叉使用时应注意其兼容性。

EPS 格式支持在文件中嵌入色彩信息 ICC 特性文件（如图 2-4 所示）。EPS 文件可嵌入两个颜色信息文件，一个是校样设置信息，另一个是 ICC 特性文件。当选择嵌入校样设置色彩信息时，图像文件的色彩信息将按校样设置中的信息进行转换，即所有图像上的色彩信息都将按校样设置的特征进行转换。当选择 ICC 特性文件时，图像文件的色彩信息仅是带有输出状态的信息，而图像的色彩信息将不会发生任何变化。

PostScript 色彩管理将文件数据转换为打印机的颜色空间。如果打算将图像放在另一个有色彩管理的文档中，请不要选择此选项。只有 PostScript Level 3 打印机支持 CMYK 图像的 PostScript 色彩管理。若要在 Level 2 打印机上使用 PostScript 色彩管理打印 CMYK 图像，请将图像转换为 Lab 模式然后再以 EPS 格式存储。

图 2-4　Photoshop 存储 EPS 格式

EPS 文件可以同时携带与文字有关的字库的全部信息。特别强调的一点是：在向非 PostScript 设备输出时，只能输出低分辨率像素。只有在 PostScript 输出设备上才能得到高分辨率的输出。所以在许多情况下，打印的校样图像非常粗糙，其原因是操作者使用的是非 PostScript 打印机来打印 PostScript 文件。只要换成 PostScript 打印机，近乎完美的图像就能打印出来。

三、便携式文档格式（PDF）及特点

便携式文档格式，也称 PDF（Portable Document Format）格式，是由 Adobe Systems 用与应用程序、操作系统、硬件无关的方式进行文件交换所发展出的文件格式。PDF 文件以 PostScript 语言图像模型为基础，无论在哪种打印机上都可保证精确的颜色和打印效果，即 PDF 格式文件会忠实地再现原稿的每一个字符、颜色以及图像。

可移植文档格式是一种电子文件格式。这种文件格式与操作系统平台无关，也就是说，PDF 文件不管是在 Windows、Unix 还是在苹果公司的 Mac OS 操作系统中都是通用的。这一特点使它成为在 Internet 上进行电子文档发行和数字化信息传播的理想文档格式。越来越多的电子图书、产品说明、公司文告、网络资料、电子邮件开始使用 PDF 格式文件。

Adobe 公司设计 PDF 文件格式的目的是支持跨平台协作、多媒体集成信息的出版和发布，尤其是提供对网络信息发布的支持。为了达到此目的，PDF 具有许多其他电子文档格式无法相比的优点。PDF 文件格式可以将文字、字型、格式、颜色及独立于设备和分辨率的图形图像等封装在一个文件中。该格式文件还可以包含超文本链接、声音和动态影像等电子信息，支持特长文件，集成度和安全可靠性都较高。

PDF 主要由三项技术组成：

（1）衍生自 PostScript，用以生成和输出图形；

（2）字型嵌入系统，可使字型随文件一起传输；

（3）结构化的存储系统，用以绑定这些元素和任何相关内容到单个文件，带有适当的数据压缩系统。

PDF 文件使用了工业标准的压缩算法，通常比 PostScript 文件小，易于传输与储存。它还是页独立的，一个 PDF 文件包含一个或多个"页"，可以单独处理各页，特别适合多处理器系统的工作。此外，一个 PDF 文件还包含文件中所使用的 PDF 格式版本，以及文件中一些重要结构的定位信息。正是由于 PDF 文件的种种优点，它逐渐成为出版业中的新宠。

PDF 文件格式在 20 世纪 90 年代早期开发，能够跨平台操作，包括文件的格式、内置图像的分享。在当时万维网及 HTML 尚未兴起之时的最初几年中，PDF 在桌面出版工作流技术当中很受欢迎。

从 Adobe PDF 2.0 版开始，Adobe 开始免费分发 PDF 的阅读软件 Acrobat Reader（现时改称 Adobe Reader，创建软件依然称为 Adobe Acrobat），而旧的格式依旧支持，使 PDF 后来成为固定格式文本业界的非正式标准。2008 年，Adobe Systems 的 PDF 参考 1.7 版成为 ISO 32000-1:2008，从此 PDF 就成为正式的国际标准。亦因为这个缘故，现时 PDF 的更新版本开发（包括未来的 PDF 2.0 版本的开发）变成由 ISO 的 TC 171 SC 2 WG 8 主导，但 Adobe 及其他相关项目的专家依然有参与其中。

最初 PDF 只被看作一种页面预览格式，而不是生产格式。Adobe 公司于 2009 年 7 月 13 日宣布，作为电子文档长期保存格式的 PDF/Archive（PDF/A）经中国国家标准化管理

委员会批准已成为正式的中国国家标准，并已于 2009 年 9 月 1 日起正式实施。PDF 格式文件成为数字化信息事实上的一个工业标准。

1. ISO 标准化

自 1995 年起，Adobe 公司参与了一些由 ISO 创建出版技术规范及在用于特定行业及用途的 PDF 标准专业子集（如 PDF/X 或 PDF/A）进程中与 ISO 协作的工作组。制定完整 PDF 规格的子集的目的是移除那些不需要或会对特定用途造成问题以及一些要求的功能的使用在完整 PDF 规格中仅仅是可选的（不是强制性的）功能。

2007 年 1 月 29 日，Adobe 公司宣布将发布完整的 PDF 1.7 规格给美国国家标准协会（ANSI）及企业内容管理协会（AIIM），为了由国际标准化组织（ISO）发布。ISO 将制定 PDF 规格是未来版本，而且 Adobe 仅仅是 ISO 技术委员会的一员。

ISO "全功能 PDF" 的标准在正式编号 ISO 32000 之下发布。全功能 PDF 规格意味着不仅仅是 Adobe PDF 规格的子集；就 ISO 32000-1 而言全功能 PDF 包含了 Adobe 的 PDF 1.7 规格定义的每一条。然而，Adobe 后来发布了不是 ISO 标准的一部分的扩展。那些也是 PDF 规格中的专有功能，只能作为额外的规格参考。

2. 历史版本

（1）PDF 1.0

PDF 1.0 于 1992 年秋季在计算机经销商博览会（COMDEX）发布，该技术获得 Best of COMDEX 奖项，用以创建和查看 PDF 档案的工具 Acrobat 于 1993 年 6 月 15 日推出，它已经有内部链接、书签和嵌入字体功能，但唯一支持的色彩空间是 RGB，对印前操作不支持。

（2）PDF 1.1

Acrobat 2 于 1994 年 9 月上市，它支持新的 PDF 1.1 档案格式，PDF 1.1 新增的特点包括：外部链接（External Link）、文章阅读器（Article Threads）、保全功能（Security Features）、设备独立色彩（Device Independent Color）和注解（Notes）。

（3）PDF 1.2

1996 年 Adobe 公司推出 Acrobat 3.0 及配套 PDF 1.2 规格，PDF 1.2 是第一个真正可用在印前环境的 PDF 版本，除了表单外，包括支持 OPI 1.3 规格、支持 CMYK 色彩空间、PDF 内包含专色定义、半色调函数（Halftone Function）与叠印（Overprint）指令等印前相关的功能。

（4）PDF 1.3

1999 年 4 月 Adobe 公司推出了在内部被称为'Stout'的 Acrobat 4，它为我们带来 PDF 1.3，新的 PDF 格式规格包括支援：双位元的 CID 字体；OPI 2.0 规格；称为 DeviceN 的一个新色彩空间，改善专色能力；平滑渐变层（smooth shading），有效率平滑渐变层的技术（从一个色彩渐变到另一个色彩）；注解（annotations）。

（5）PDF 1.4

2000 年中 Adobe 公司推出 Illustrator 9，它是第一个支持 PDF 1.4 及其透明度特征的应用程序。这是第一次 Adobe 公司并未伴随新版本的 PDF 规格而推出一个新版本的 Acrobat，虽然 Technote 5407 记载了 PDF 1.4 支持透明度，但他们也没有释放 PDF 1.4 的全部规格。

（6）PDF 1.5

2003 年 4 月 Adobe 公司宣布 Acrobat 6 于同年 5 月下旬开始使用，新版本的 Acrobat 同时带出了一个新版本的 PDF 格式，即 PDF 1.5。

PDF 1.5 有一些新的功能：

①改良的压缩技术，包括对象流（Object Stream）与 JPEG 2000 压缩；

②支持层（Layers）；

③提高标签（Tagged）PDF 格式的支持；

④ Acrobat 软件本身提供更多于新的 PDF 档案格式的立即好处。

（7）PDF 1.6 及 PDF1.7

2005 年 1 月 Adobe 推出具有 PDF 功能的 Acrobat 7，PDF 1.6 提供了下列改进：

①改进的加密演算法；

②注解和标注功能的一些小改进；

③ OpenType 字体可直接嵌入 PDF，不再需要以 TrueType 或 PostScript Type 1 字体型式嵌入。

Adobe Acrobat 8.0 于 2006 年 11 月面世，但是它不使用 PDF 1.7 作为预设的档案格式，而是使用 PDF 1.6。对印刷和印前作业而言，PDF 1.3 或 PDF 1.4 就足够了；其他的新功能包括改良的 PDF/A，更好的选单与工具的组合，能在 Adobe Reader 8.0 内储存表格，预检引擎能处理多项更正（称为 Fix-ups）。令大多数人认可，性能提高，特别是对 Intel Mac 电脑版的性能改善最多。PDF 1.7 在 2008 年 1 月成为正式的 ISO 标准（ISO 32000）。

第五节 字体技术

一、字库的类型

字库是外文字体、中文字体以及相关字符的电子文字字体集合库，被广泛用于计算机、网络及相关电子产品上。在平版制版中常用 TrueType 字库和 PostScript（PS）字库。

1. TrueType 字库

TrueType（以下简称 TT）是由美国 Apple 公司和 Microsoft 公司联合提出的一种新型数字字形描述技术。

TT 是一种彩色数字函数描述字体轮廓外形的一套内容丰富的指令集合，这些指令中包括字形构造、颜色填充、数字描述函数、流程条件控制、栅格处理器（TT 处理器）控制，附加提示信息控制等指令。

TT 采用几何学中的二次 B 样条曲线及直线来描述字体的外形轮廓，二次 B 样条曲线具有一阶连续性和正切连续性。抛物线可由二次 B 样条曲线来精确表示，更为复杂的字体外形可用 B 样长曲线的数学特性以数条相接的二次 B 样条曲线及直线来表示。描述 TT 字体的文件（内含 TT 字体描述信息、指令集、各种标记表格等）可能通用于 Mac 和 PC 平台。在 Mac 平台上，它以"Sfnt"资源的形式存放，在 Windows 系统上以 TTF 文件出现。为保证 TT 的跨平台兼容性，字体文件的数据格式采用 Motorola 式数据结构（高位在前，低位在后）存放。所有 Intel 平台的 TT 解释器在执行之前，只要进行适当的预处理即可。Windows 的 TT 解释器已包含在其 GDI（图形设备接口）中，所以任何 Windows 支持的输出设备，都能用 TT 字体输出。

TT 技术具有以下优势。

（1）真正的所见即所得

由于 TT 支持几乎所有的输出设备，所以对于目标输出设备而言，无论系统的屏幕、激光打印机或激光照排机，所有安装了 TT 字体的操作系统均能在输出设备上以指定的分辨率输出，所以多数排版类应用程序可以根据当前目标输出设备的分辨率等参数，对页面进行精确的布局。

（2）支持字体嵌入技术

支持字体嵌入技术，保证文件的跨系统传递性。TT 嵌入技术解决了跨系统间的文件和字体的一致性问题。在应用程序中，存盘的文件可将文件中使用的所有 TT 字体采用嵌入方式一并存入文件，使整个文件及其所使用的字体可方便地传递到其他计算机的同一系统中使用。字体嵌入技术保证了接收该文件的计算机即使未安装所传送文件使用的字体，也可通过装载随文件一同嵌入的 TT 字体来对文件进行保持原格式，使用原字体的打印和修改。

（3）操作系统平台的兼容性

目前 Mac 和 Windows 系统均提供系统级的 TT 支持。所以在不同操作系统平台间的同名应用程序文件有跨平台兼容性。如在 Mac 计算机上的 PageMaker 可以使用已安装了文件中所用的所有 TT 字体，则该文件在 Mac 系统上产生的最终输出效果将与在 Windows 下的输出效果保持高度一致。

（4）ABC 字宽值

在 TT 字体中的每个字符都有其各自的字宽值，TT 所用的字宽描述方法比传统的 PS 所用的更真实。TT 解释器已包含在其 GDI（图形设备接口）中，所以任何 Windows 支持的输出设备，都能用 TT 字体输出。

在 Windows 中，系统使用得最多的就是 *.TTF（TrueType）轮廓字库文件，它既能显示也能打印，在任何情况下都不会出现锯齿问题。而 *.FOT 则是与 *.TTF 文件对应的字体资源文件，它是 TTF 字体文件的资源指针，指明系统所使用的 TTF 文件的具体位置，而不用必须指定到 FONTS 文件夹中。*.FNT（矢量字库）和 *.FON（显示字库）的应用范围都比较广泛。

2. PostScript 字库

PostScript（简称 PS）是由 Adobe 公司在从前的一种面向三维图形的语言基础上重新整理制作，于 1985 年发布的页面描述语言，它是桌面系统向照排设备输出的界面语言，专门为描述图像及文字而设计。PS 的作用是将页面上的图像文字用数字公式的方法记录及在计算机上运行，最后通过 PostScript 解码器翻译成所需的输出，比如显示在屏幕上，或在打印机上运行，最后通过 PostScript 解码器翻译成所需的输出，比如显示在屏幕上，或在打印机、激光照排机上输出。

PostScript 语言是目前国际上最流行的页面描述语言形式，它拥有大量可以任意组合使用的图形算符，可以对文字、几何图形和外部输入的图形进行描述和处理，从理论上来说可以描述任意复杂的版面。其丰富的图形功能、高效率地描述复杂版面的功能，吸引了众多出版系统的排版软件和图形软件对它的支持，几乎所有的印前输出设备都支持 PS 语言，而 PS 语言的成功，也使开放式的电子出版系统在国际上广泛流行。

20 世纪 80 年代末 PS 语言成为事实的行业标准。经过多年经验的积累和许多 PS 产品的反馈，1990 年推出 PS2，在 1990 年进而推出 PS3。PostScript 字库技术经历了开始的 Type 1、Type 3 格式，1990 年复合字库 Type 0 格式（OCF）发表。

二、前端字体与后端字体

前端字库指的是显示字库，当您在排版软件中使用这些字体时，可以显示出字体的实际效果；后端字库是给输出软件用的，也就是我们常说的 CID 字库，如果排版软件在生成 PS 文件时未下载字体，发排软件输出时会调用 CID 字库，后端字库要比前端字库输出的质量好。

前端字体就是 TrueType 字体，用于显示和打印。TrueType 字体都安装在系统的字体文件夹中，在排版软件中使用。后端字体是指后端 RIP 中使用的字体，一般常见的是 CID 字体。当后端 RIP 中的字库涵盖前端字体时，前端排版软件在生成结果文件时不用下载使用过的字体。

1. TrueType 前端显示字库

TrueType 前端显示字库指装在排版主机上用于屏幕显示的字库（显示字库在飞腾、书版、Word 等软件上都可以使用，但非方正软件不能使用 748 码字库）。飞腾排版时使用的字库即为显示字库，常见的有 748 码、GB、GBK、BIG5、超大字库集等。

（1）方正兰亭。可用于 Windows 95/2000 平台上的标准 TrueType 字库，适用于 Windows 平台上的所有通用软件和方正软件，提供 GB、BIG5 和 GBK 三种编码。

（2）方正妙手。可运用于 Mac 平台上的 Mac TrueType 字体，适用于 Mac OS 平台上所有通用软件，提供 GB 和 BIG5 两种编码。

在 PC 平台上可以安装方正兰亭字库进行排版设计，在 Mac 平台上可以安装方正妙手字库进行排版设计。

2. PostScript 后端发排字库

PostScript 后端发排字库指安装在后端输出设备（如照排机、打印机）中用于发排的字库，也称 PS 字库（这种字库安装在后端 RIP 软件上，如 PSPPRO、PSPNT）。发排字库不能在屏幕上显示。方正发排字库按其编码的不同可以分为 748 码、GB、GBK、BIG5、超大字库集等。

方正发排字库与方正显示字库是一一对应的。

（1）方正文韵。可以安装在 PSPNT 上的 PostScript Type 0 字库，从 PC 或 Mac 上直接安装，提供 748 码、GB、BIG5 以及 GBK 四种编码，当连接 PSPNT 的输出设备超过 1450 dpi 时，不能使用方正文韵字库进行输出。

（2）方正天舒。和方正文韵格式一样，但可以在 1450 dpi 以上的设备上输出。在安装完方正文韵字库或方正天舒字库后，在 PSPNT 的目录下会有 Fonts 和 Fzdata 两个子目录。

（3）方正 CID 字库。PSPNT 的专用字库，提供 748 码、GB、BIG5 以及 GBK 四种编码格式，并提供一套超大字库。方正 CID 字库缺省安装在 PSPNT 的 Font 目录下。

如果前端排版软件使用了方正兰亭和方正妙手字库时，PSPNT 上没有安装相应兰亭字库，并在"重置"对话框选择"Windows 系统的 TrueType 字体"选项，就可以输出。同样，也可以在 PSPNT 主机上安装汉仪 TrueType 字库进行输出。但要注意如果在前端排版软件（如 QuarkXpress、FreeHand）中对字体做了变形效果，则 TrueType 字或 CID 字输出会报语法错误。

在 PSPNT 中进行重置字库时，会有一个字库识别的顺序问题，当选择了"使用 Windows 系统的 TrueType 字体"选项时，首先识别 Windows 系统下的 TrueType 字，然后识别"字库路径"中指定的方正 CID 和 Type1 字库，如果安装了方正天舒、方正文韵以及汉仪 PostScript Type0 等第三方字库，那么最后会识别第三方字库，即"PSPNT/Fonts"下的字库。当这三种格式的字库有重名时，使用最后识别出的字库进行输出。

第六节 拼版的基本概念与方法

在工作中不会总是做 16 开、8 开等常规开本的产品，特别是包装盒、小卡片（合格证）等产品常常是不合开的，这就需要在拼版的时候注意尽可能把成品放在合适的纸张开度范围内，以节约成本。

一、常规的拼版

根据印刷的需要（如数量）以及设备的限制，如 8 开机、4 开机、对开机、全张机等，拼版的时候也要按实际情况进行不同的调整，一般拼 8 开或 4 开就足够用了，因为在对开和全开的印刷机上可以用套晒、拼晒，并通过自翻身或正反印来解决。

1. 单页形式的印刷品

拼版时中间（垂直中线）拼接部分留 6 mm 出血，即每个单页四边均留 3 mm 出血（需要切两刀）。需要注意，如果你做的产品没有出血的图片、底纹，或完全是一色底纹等，可以按单页形式的方法拼版，中间一刀即可。

2. 封套的拼版

一般制作的时候，习惯把封套连同"舌头"拼在一起，这种做法比较费纸（有一块空白没有利用），但图案连续性好。还有一种方法是封面归封面，"舌头"单独做，这样做省纸，但多一道"糊工"，即在成品时多刮一次胶（或多贴一道双面胶带）。

3. 包装盒的拼版方式

一般大包装盒（超过 8 开的）不用拼版，直接套晒就可以了。

4. 简单小包装的拼版

尽量在合开的前提下，把拼版工作做到最紧凑，但包装盒牵涉的后道工艺比较多，轧盒（切出边缘并压折痕线）是最关键的，这时需要注意拼版时最近的两条边线间应不小于 3mm，否则在做刀模的时候会很麻烦甚至影响产品质量。

注意问题：

（1）根据活件和印刷机尺寸确定拼板尺寸；

（2）手工拼版注意页码的位序及正反方向；

（3）留出血位，一般为 3～5 mm；

（4）每一个单面页码和文字的对齐，才能保证精度；

（5）不同克重的纸张经过折页页码会有一定的偏移，应处理好爬移量；

（6）157g 纸张以上不建议三手折页（即便是开花线）。

二、滚翻印版编辑

滚翻印刷是指一个印版纸张两面各印一次，印完一面后，纸张翻面旋转 180°，再印第二面。印第二面时，纸张的叼口方向要改变。印后沿中间裁切后可得两份同样的印刷品，如图 2-5 所示。这种方法适用于印数不多、一个印版上放有印刷品正反两面内容、印刷机幅面相对较大等情况。例如，要印刷一个产品广告说明书折页，印刷成品尺寸为 87 mm × 180 mm 的 6 折页，准备在四开机上印刷。全开正度纸的尺寸为 787 mm × 1092 mm，四开纸的尺寸约为 540 mm × 390 mm，6 折摊开为 522 mm × 180 mm。这样，一张四开纸上可放 522 mm × 180 mm 的两个产品说明书折页。这样可以把正反面内容拼在一个 4 开版上，印刷时采用滚翻版印刷，裁切后，一个 4 开可得两个说明书。

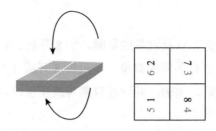

图 2-5 滚翻印版（前后翻版）

三、自翻印版

自翻印刷是一块印版在纸两面印刷，但纸的翻法是常规翻法，叼口方向不变，如图 2-6 所示。一般 16 开杂志封面印刷常采用这种方式。例如，印一种杂志的 4 个封面，可以把版拼成一个 4 开，然后上 4 开机印刷。印完一面后自翻印刷，裁切后一个 4 开可得两个封面。

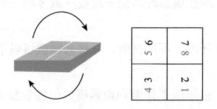

图 2-6 自翻印版（左右翻版）

第七节 CTP 加网技术

传统的网目调加网就是人们常说的调幅加网，它由一系列规则排列的网点组成，通过网点尺寸的变化来体现高光或暗调的区别。调幅加网具有很强的预测性，而且在印刷机上的效果比随机加网要好。这种技术的局限性就在于，印刷厂必须保证图像在最亮和最暗的地方不丢网点。随着加网线数的增加，控制高光网点和保持暗调网目间的间隙就变得越来越困难了。

调频加网或随机加网使用了与调幅加网完全不同的方式，它们能够保持网点尺寸的一致，但会改变网点之间的距离。由于在很多区域内网点非常小而且排列得非常紧密，所以调频加网能够表现出更多的细节（中间调），同时减少莫尔条纹出现的机会。用调频加网技术印刷出的图像往往能够达到连续调照片的质量。但是，由于调频加网使用了非常小的网点，所以在调幅加网过程中，常常出现的高光网点的问题也会出现在调频加网的大部分阶调中。在计算机直接制版技术出现以前，人们很难将微网点从胶片上转移到印版上，而现在，由于新技术的出现这一问题得以解决。

很多新兴的加网技术都是调幅和调频加网的混合体。混合加网在大部分色调区域内使用传统的调幅加网方式，而在高光和暗调区域使用调频加网技术。通过混合加网，平版胶印厂能够提高调幅加网区域的加网线数，而不会给印刷机带来额外的负担。以下就是目前市场上比较常见的几种最新加网产品（爱克发、艾司科、富士胶片、海德堡、柯达、网屏）。

一、爱克发

爱克发公司声称自己推出的加网技术能够将调幅和调频加网的优势结合起来。该公司的 Sublima 就是一个将两种技术整合到一个解决方案中的专利产品。它同时是爱克发公司第一个采用 XM 专利技术的解决方案。在 Sublima 的开发过程中共使用了两项爱克发技术，它们分别是 ABS （Agfa Balanced Screening）和 Cristal Raster 调频加网技术。

Sublima 的工作方式，在中间调区域，Sublima 使用 ABS 技术进行清晰准确的复制，在高光和暗调区域，用调频技术来再现图像的细节内容。从一种技术转换到另外一种是使用专利技术来确定两种加网技术之间的切换点，并实现平稳过渡，不会影响图像的效果。虽然调频加网区域使用了比较小的网点，但它们还是会按照 ABS 建立的加网角度进行排列，这样做的最终结果是生成一个全新的网目。Sublima 所能达到的加网线数分别为 210 lpi、240 lpi、280 lpi 和 340 lpi。

Sublima 软件能计算出印版在每一种印刷机上所能印刷出的最小的网点尺寸，允许印刷工人在印刷机上进行调整——这点它优于调频加网。

Sublima 内置的校准曲线能够自动补偿不同的网点增大。它能在长版印刷中保持 1% 到 99% 的网点不丢失。这项技术最大的优势还是体现在印前和印刷方面，在最高的加网线数下，在 2400dpi 的分辨率进行 RIP，但是 340 lpi 的印刷效果与 150 lpi 的印刷效果没有任何区别。

二、艾司科

在 2006 年 9 月，艾司科公司推出了全新的 Perfect Highlights 柔性版印刷加网技术。它为艾司科用户、纸盒加工厂和标签印刷厂带来了全新的工具，使他们知道如何在丝网印刷过程和制版过程中达到最好的效果。印刷厂能够为特定的印刷环境、油墨、承印物和印刷机设定最佳的加网参数，提高包装购买者所获得的价值，为他们带来更大的竞争优势。Perfect Highlights 能通过各种方式印刷出 1% ～ 2% 的高光网点，与艾司科其他加网技术一起为用户带来最佳的中间调和暗调复制效果。

此外，艾司科公司还开发出了包括 SambaFlex 和 Groovy Screens 在内的多种加网技术。SambaFlex 和 Groovy Screens 都是混合加网技术，它们会在图像的大部分区域使用调幅网点，并在高光和暗调部分的线条区域使用调频网点，以提高色彩的饱和度和图像的立体效果。艾司科公司还拥有 HighLine 调幅加网技术，它能够在较低的输出分辨率下生产出高加网线数的产品。如能够在 2400 dpi 的分辨率下生产出 423 lpi 的网点。

三、富士胶片

富士胶片公司拥有两项先进的加网技术——Co-Res AM Screening（针对普通分辨率的图像）和 Taffeta FM Screening。

Co-Res Screening 是由富士胶片公司开发的一款革命性高精度加网软件，它能让客户在使用标准输出分辨率的情况下印刷出高加网线数的图像。这样一来，用户就能提高高线数加网的生产力，同时在高光区域得到更加精细的复制效果。

富士胶片公司开发出的 Taffeta 结合了富士胶片公司特别设计的功能完善的数字成像技术 Image Intelligence，能够帮助印刷厂解决调频加网过程中常常出现的问题，并能有效增加彩色复制的范围和精度，减少莫尔条纹的产生。新型 Taffeta FM Screening 采用了颗粒优化算法，用光学特性来模拟印刷效果，并用网点形状优化算法来减少图像的颗粒感，提高印版的印刷适性。

富士公司声称 Taffeta 具有以下特点：
（1）完全消除莫尔条纹和玫瑰斑；
（2）提高基本色和间色的饱和度；
（3）更好地再现图像细节；
（4）改善印版的质地和印刷适性；
（5）消除印版的不均匀和颗粒感。

四、海德堡

海德堡公司在 2006 年 4 月向英国市场推出了新的 Prinect 混合加网方法。这个新型加网方法能将调幅和调频加网技术的优点结合起来。它能为用户带来更高的加网分辨率和更清晰的细节内容，从而提高图像的整体印刷质量。Prinect 混合加网技术已经被用在了 Suprasetter 系列热敏制版机、Prosetter 系列紫激光制版机以及 Topsetter 系列制版机上。根据 CTP 制版技术和印版类型的不同，它最高能达到 400 lpi 的加网分辨率。

Prinect 混合加网技术能为高光和暗调区域定义出最小的网点，而且不允许人们使用低于这个最小尺寸的网点。与随机调频加网不同的是，这项技术所使用的网点将按照调幅加网的角度排列。这就能保证不同阶调值之间的平稳过渡。此外，Prinect 混合加网技术将混合网点分布在相关的角度上，以增强图像细节部分的清晰度，减少莫尔条纹的产生。Prinect 混合加网系统采用了海德堡自己的 Irrational Screening（IS）技术，这不但能有效消除莫尔条纹，而且能在比较难复制的皮肤区域实现更加平滑的效果。对于黑白印刷品，这个系统还能对黑色的加网角度进行单独设置，从而达到更好的效果。

海德堡公司升级的 Satin Screening 调频加网技术整合到 Prinect 工作流程中，形成 Prinect 随机加网技术。

五、柯达

柯达 Staccato 软件是第二代调频混合加网产品，它能为用户带来更自然的阶调和更平滑的色彩，它非常适合用在平版胶印产品中。Staccato 加网软件能够生产出连续调的高保真图像，这些图像的细节清晰、色域广、质量高，完全不会出现莫尔条纹和玫瑰斑等问题。Staccato 软件所能提供的网点尺寸在 10 μm 到 70 μm 之间。

据柯达公司介绍，Staccato 加网软件消除灰度的限制和不同色调之间生硬的衔接，同时提高颜色和网目调的稳定性。Staccato 还能减少印刷机套准误差所造成的图像变形问题。通过使用一个 10000 dpi 的激光，柯达 Sqaurespot 热敏成像技术能够为用户带来 Staccato 加网软件所需的分辨率，从而帮助人们在日常生产中复制出更加稳定和精细的网点。Staccato 为四原色印刷提供四种不同的网点形状方，还提供另外六种网点形状的选择，以支持柯达 Spotless 印刷技术的应用。

六、网屏

网屏公司发布了自己最新的混合加网系统视必达（Spekta 2 HR）。Spekta 2 既能达到调频加网精美的印刷质量，又能实现调幅加网稳定的印刷效果，能够为用户带来与 350 lpi 调幅加网的图像水平相当的印刷质量和细节清晰度。据网屏公司介绍，Spekta 2 能够很轻松地调整网点扩大的问题，因为它使用了网屏公司独有的 12 bit 加网技术，Spekta 2 提高了人们对色彩的整体把控能力，并且保证了图像的清晰度和平滑度。Spekta 2 HR 对细节的再现能力能够与 650 lpi 以上的加网线数相媲美。

第八节　印刷补偿曲线和反补偿曲线基本知识

实际生产中要求 CTP 印版能够线性输出，使印版的网点大小接近电子文件数据，保证网点转移的准确性。但实际上，印版输出时如果不做任何补偿，是不能实现线性输出的，实际曲线是使用没有进行印刷补偿的印版印刷后，测量印品的网点梯尺获得的网点曲线，而目标曲线则希望得到印品网点增大曲线。为了达到目标曲线需要在输出印版时对其进行相应补偿。获得补偿值的方法与印版线性化基本相同。

一、印刷补偿曲线

CTP 线性化目的是建立数字印版文件的网点与印版上网点的关系，以控制最终曝光后印版上网点的大小。无线性化时的文件，即转换曲线为 45°的一条直线。流程软件解释输出时将按照数字文件网点值进行解释，从而保证数字文件的网点值与印版上的网点值大小一致。从图 2-7 可以看出，CTP 设备在不做补偿时的输出是非线性的，因此需要确定特定曝光与显影条件下印版的输入数据，使印版输出达到线性输出的条件。

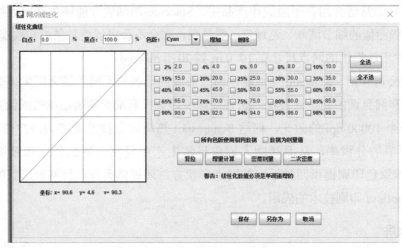

图 2-7　不做调整的线性转换曲线

（1）线性化调整的目的：使印刷品的网点大小接近原稿。

（2）线性化调整的原因：整个印前及印刷流程里有很多影响网点大小的因素，如晒版的时候由于胶片和 PS 版之间有空隙，曝光时网点会有变化；印刷机在印刷的时候由于压力的作用会使网点扩大等。

（3）线性化调整的方法：印版检测、印刷品检测等。

（4）线性化调整的工具：网屏 CTP 的 SGD 或者 DotGain 等。

二、建立印刷补偿曲线的步骤

1. 印版的测量

（1）用 Illustrator 图形处理软件制作 5mm×6mm 包含 30 块 0%～100% 的色块图，在流程软件中选用圆形网点（即方圆形）以 2540 dpi、175 lpi 不加任何线性曲线；

（2）在流程中设定相应的参数：线性化曲线 -NONE，微调曲线 -NONE，印刷补偿曲线 -NONE，印刷反补偿曲线 -NONE；

（3）加网解释，在 CTP 版材上输出单色版；

（4）使用印版测量仪测量该版材中 50% 处的 9 个色块的网点面积率，以判断版材上网点的均匀性，要求平均误差 ≤ 2%；

（5）测量 0%～100% 各色块，得到第一次印版的网点面积率，计算它们的平均值。

2. 制作线性化曲线

在流程中选取"输出设备"→"曲线管理"→"PDF 加网"命令，输入在印版上各色块所测得的网点面积平均值。输入时要特别注意线性化曲线的光滑度，不要有突变，必要时可以忽略个别控制点。

得到的印版线性文件加到"线性化曲线"之中，并再次加网输出印版测量，以验证曲线是否符合要求。若没有达到要求，可以循环几次以得到满意的印版线性化曲线。最终实测值与需要的值基本一致，如在 50% 的网点处，印版实测输出值为 50%±1% 即可。

将需要印刷的测试文件应用之前得到的线性化曲线在工作流程中加网并输出印版。

3. 制作印刷补偿曲线

（1）用 Adobe Illustrator 图形处理软件制作色阶梯尺，网点百分比值见表 2-1 的梯尺网点百分比所对应的列的数值；

表 2-1 梯尺网点百分比

单位：%

梯尺网点百分比	期望印品网点百分比（ISO 标准）	印版测量网点百分比
0	0	0
2	3.1	6
4	6.3	11
6	9.4	16
10	15.7	23
20	30.5	40
30	44.3	55
40	56.9	66
50	68.4	76
60	78.3	83
70	86.5	89
80	93	94
90	97.5	98
96	99.3	100
97	99.5	100
98	99.7	100
99	99.8	100
100	100	100

（2）根据 ISO 印刷标准确定期望的印品网点百分比值（如表 2-1 第二列所示，以 ISO 12647-2：2016 为标准）；

（3）测量输出后印版上对应色阶上的网点百分比值（如表 2-1 第三列所示）；

（4）绘制网点阶调曲线，如图 2-8，并推算线性补偿各阶调的网点百分比值。下面以梯尺 50% 的网点处为例，介绍一下线性化补偿曲线的生成原理。

①从 50% 处作垂线，与期望值曲线相交于 A 点；

②从 A 点做水平线，与当前设备复制曲线相交于 B 点；

③从 B 点作垂线与 45°斜线相交于 C 点；

④从 C 点做水平线，与 A 和 x 轴 50% 处连线相交于 D 点。则 D 点即为在梯尺网点 50% 处要得到期望值曲线上 A 点所需要的校正参数。同理，可得一系列校正点，将所有校正点全部连起来，构成的曲线叫线性化补偿曲线（图 2-8 中的虚线）。

在输出 CTP 版时，选择线性化补偿曲线，就可使输出最终印品和 ISO 标准达到一致。

在建立和使用线性化时需要注意以下事项：

①建立线性化时，四套色版的线性化数据可能存在差异，这种差异随着加网线数的提高而逐渐变大。在输出高线数印版时，就需要单独为每一个色版分色做线性化；

②加网线数在 175 lpi 与 200 lpi 时，可以使用单色印版的线性化文件替代分别制作四色印版的线性化文件，此时要选择"所有色版使用相同数据"，使其生成曲线来代表所有的色版；

③印刷机补偿曲线目的是要找到印刷机的网点扩大特性，应根据目标网点扩大数据生成补偿曲线。如 175 lpi 方圆网点在 50% 网点处的扩大率一般控制在 15% 左右。

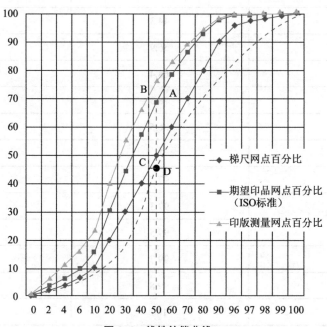

图 2-8 线性补偿曲线

第三章 印版输出准备工作

学习目标

1. 掌握平版制版生产作业的环境要求；
2. 能选择输出线性化曲线；
3. 能导出便携式文档格式（PDF）、1位TIFF和8位TIFF；
4. 能使用工作流程转换颜色；
5. 能使用工作流程软件拼版；
6. 能在激光照排机及直接制版机上装、卸印版；
7. 能设置印版尺寸；
8. 能启动机器并操作控制面板；
9. 掌握显影机的维护、保养方法。

第一节 生产作业的环境要求

平版制版生产作业的环境包含两部分：一部分是流程输出环境，另一部分是CTP设备输出环境。

一、流程输出环境

1. 服务器（以科雷印艺汇通为例）

一台服务器上安装印艺汇通服务器程序。用户可见的是一个名为"流程服务器"的程序，它的功能是用来启动及关闭印艺汇通服务器。

服务器上安装数据库系统。印艺汇通服务器程序使用该数据库存储及管理所有作业运行信息。数据库系统一般随着计算机系统的启动而自动启动。

服务器上存储工作数据。所有的原始页面文件以及作业运行中间产生的永久的或临时的数据文件都存在流程服务器上。

2. 流程工作站

若干台流程工作站上安装有印艺汇通客户端程序。用户主要通过流程客户端软件进行日常生产工作，这些工作包括创建作业、收集源文件、设计折手方案、数码打样、输出印版，等等。

客户端程序中内置折手版式设计程序。用户不能直接启动折手版式设计程序，只能在流程客户端中通过新建版式或编辑修改版式来启动该程序。

客户端程序中还内置自由拼版程序。用户不能直接启动自由拼版程序，只能在流程客户端中通过建立一个自由拼版模板，并将待拼版的散页文件提交给它，来启动自由拼版程序工作。

二、CTP 设备输出环境

（1）CTP 设备工作间选址要求：最好不要选择上下楼层或附近存在大型设备，因大型设备工作时的噪声或振荡可能影响到 CTP 设备，具体接线如图 3-1 所示。

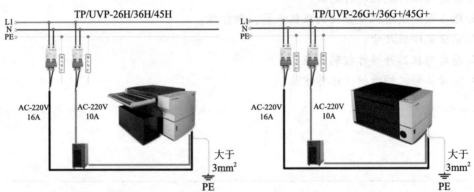

图 3-1 CTP 制版机接线

（2）CTP 工作间的空气指标，要求达到国家质量二级标准，即 API 值大于 50 且小于等于 100。

（3）API 值指的是空气污染指数，达到 50 是一级标准，达到 100 是二级标准。

（4）恒温恒湿的作业场地，最佳温度要求 23℃ ±2℃，最佳湿度要求 50%±10%。

（5）作业场地水平要求在 4 mm 以下。

（6）电源的要求是单相线不能少于 4 mm^2，单相线空开不得低于 25A。

（7）对地电阻要求是采用直径 4 mm^2 地线连接到直径不小于 10 mm 的镀锌金属棒，

金属棒要深埋湿土 1.5 m 或以下，设备对地电阻低于 0.5 Ω。

（8）对用于冲版设备的水质要求是提供足够的水压，如果不够需加装增压泵；对水质较差的区域，要安装过滤装置。

第二节　输出流程软件与设备驱动软件的安装

一、CTP 输出流程软件安装（以新版 Laboo 5.X 输出器为例）

安装时首先把软件安装光盘放入计算机光驱，这时会弹出安装界面，可以单独安装运行。为了能顺利安装，请在安装该软件前，将各类杀毒软件及安全卫士退出。根据提示设置语言，单击下一步，完成即可。

二、CTP 设备输出驱动软件的安装

（1）开启 CTP 电源，连接 USB 连接线到计算机 USB 端口。需要注意，开启电源之前请确认已经将 CTP 运输时采用的所有固定支架去除。

（2）单击计算机右键，选择"属性"。

（3）单击进入"设备管理器"。

安装过程如图 3-2 所示。

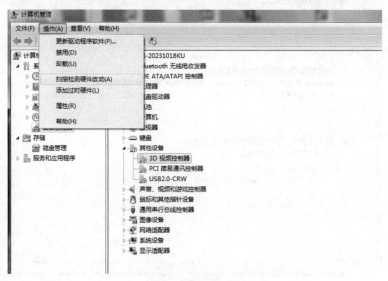

图 3-2　安装过程

（4）单击"操作"，选择"扫描监测硬件改动"进行扫描。

（5）找到连接后新显示的未安装的 USB 驱动之后，单击右键，选择"更新驱动程序软件"，如图 3-3 所示。

图 3-3　更新驱动程序软件

（6）请选择"浏览计算机以查找驱动程序软件"手动选择路径，不推荐选择自动搜索，如图 3-4 所示。

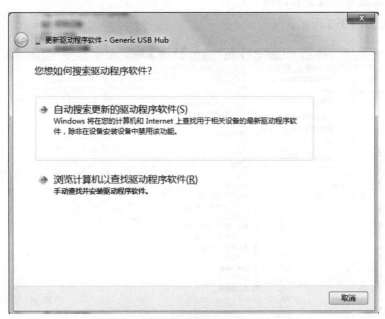

图 3-4　浏览计算机以查找驱动程序软件

（7）手动选择 USB 驱动安装路径，如图 3-5 所示，安装 WIN7 64bit 的 USB 驱动；

在 Laboo 安装目录下找到 USBDrv128B\USBDrv128B\sign Drv\win7_amd64，单击"确定"。确认路径正确之后，选择"下一步"。

图 3–5　安装 WIN7 64bit 的 USB 驱动

（8）弹出 Windows 安全对话框，选择"始终安装此驱动程序软件"。安装 USB 驱动，显示已成功安装完成 USB 驱动，单击"关闭"，如图 3-6 所示。

图 3–6　弹出 Windows 安全对话框

（9）安装完成之后，设备管理器的"通用串行总线控制器"下的第一行显示：Cron Laser device,V1C，说明 CTP 设备已成功安装 USB 驱动，如图 3-7 所示。

图 3–7　设备管理器的"通用串行总线控制器"

特别说明：如果连接设备的计算机主机是 intel 8 系列主板，则需要安装 PCI 转接口进行转换，否则可能会报"数据传输错误"，影响设备正常出版。

第三节　印版输出的线性化

使用传统制版流程，改用 CTP 直接制版流程，会发现 CTP 印品比传统印品偏暗（印刷条件不变）。这是由于传统制版流程中，数据文件首先经过照排机输出胶片，网点从胶片到阳图型 PS 版，由于晒版时光线的斜射和网点边缘的不实等因素，印版上的网点面积较胶片上的网点面积要小，再经过印刷的网点扩大；而 CTP 制版流程中没有胶片，印版上的网点面积和数据文件的网点面积一样大。因此，通过 CTP 制版机的线性化校准使印品质量与传统印刷一致或者更好。

一、创建校正曲线

创建流程中所用到的校正曲线使用资源库的"校正曲线"标签页，如图 3-8 所示。

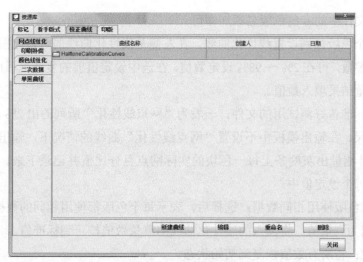

图3-8 校正曲线标签

单击"校正曲线"标签页，切换到校正曲线资源，包括网点线性化、印刷补偿、颜色线性化、二次数据、单黑曲线。其中网点线性化和印刷补偿在输出模板中使用，颜色线性化、二次数据和单黑曲线在打样模板中使用。

在"校正曲线"窗口中，可通过单击左侧的标签页来切换每种曲线的界面。在每个界面上，都可对曲线进行新建、编辑、重命名和删除。

单击"新建曲线"按钮，弹出"网点线性化"对话框，如图3-9所示。

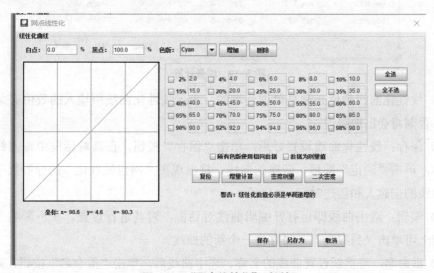

图3-9 "网点线性化"对话框

"白点""黑点"缺省值分别为0和100。除非特殊应用，一般不要改变。若白点填w%，黑点填k%，则页面中黑度为0%（空白）的地方将输出w%的网点；黑度为100%（实地）的地方将输出k%的网点，中间部分按线性化曲线调整。

"色版"默认情况下的色版为cyan、magenta、yellow、black。可以通过单击"增加"或"删除"按钮来添加或删除一个专色色版。单击"增加"按钮后，弹出"添加专色版"

对话框，根据文件中的专色色版名添加即可。

可以对每个色版分别设置线性化曲线。在左侧是线性化曲线的可视化图，右侧是设置线性化曲线的参数，可在2%～98%设定数值，在选中设定值前的复选框后所设数值生效，通常需根据测量结果填入数值。

测量方法：准备好测试用的文件，一般为"网点线性化"所列的由2%～98%的不同等级的灰度梯尺。在输出模板中不设置"网点线性化"曲线的情况下，输出该文件。用透射/反射密度计测量出灰度条上每一色块的实际网点百分比值并记录下来。将刚才的测量结果依次填入每个设定值中。

（1）所有色版使用相同数据：选择后，表示每个色版都使用相同的数据。

（2）数据为测量值：将输入的线性化曲线的测量数值校正到标准值。

（3）复位：将所有数值恢复到初始状态。

（4）增量计算：对线性化曲线的增量计算。输入在当前曲线基础上的测量值，对曲线进行增量调整。单击后，弹出"增量计算"对话框，如图3-10所示。

图3-10 增量计算

（5）线性化曲线数值必须是单调递增的：表示线性化曲线所输入的数值必须是单调递增的，否则将会出现错误。

（6）保存：线性化曲线设置好后，单击"保存"按钮，在其对话框中输入线性化曲线的名称，单击"确定"即可。新建的曲线，将出现在"网点线性化"的标签页中，可查看每条曲线的创建人和创建时间。

（7）编辑：双击曲线即可打开编辑曲线对话框，对其进行设置。在不影响上一个曲线的基础上可单击"另存为"，另存为一个新的曲线。

（8）重命名：可重新设置曲线的名称。选中曲线后，单击"重命名"按钮，弹出"重命名"对话框，设置名称即可。

（9）删除：删除选中的曲线。

二、流程软件中实现线性化

在"曲线校正"设置标签页中，可设置网点线性化和印刷补偿两种校正曲线。这两种曲线在资源库中，可根据不同的需求进行创建。

1. 网点线性化

线性化曲线用来校正输出时因设备及环境因素产生的误差。通常情况下，在使用相同的输出设备时，处于不同分辨率，用不同网线输出时，线性化曲线也会有所不同，最好为每个分辨率各建一条线性化曲线。

选中"网点线性化"后，单击"浏览"按钮，弹出"CURVES选择"对话框，如图3-11所示。

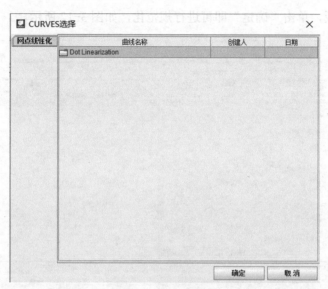

图3-11　网点线性化

在对话框中，会显示用户在资源库中创建的网点线性化曲线资源，选中所需曲线，单击"确定"即可。

2. 印刷补偿

印刷补偿用来对印刷压力所造成的网点增大进行补偿，印刷补偿曲线的设置方法与"网点线性化"设置方法相同。

第四节　输出流程软件操作

一、规范化文件

1. 提交源文件给规范化处理器

把某个作业中的一个任务指派给某个处理器节点处理的过程称为"提交"。当把一个作业"提交"给某个节点模板，就意味着将相关的页面数据以及模板中的参数一起交给服务器上相应的工单处理器（JTP）进行处理。处理的结果将显示在客户端中，处理产生的新的数据会存储在服务器上相应作业的文件夹中。

文件添加成功后，第一步，就是对文件进行规范化，即将源文件规范成标准的单个PDF文件。

选中已添加就绪的文件，单击"提交"按钮或选中文件右击，在弹出的对话框中选择"提交"选项，打开"提交"对话框。在对话框中列出了源文件可以提交的模板，选中要提交的规范化模板，单击"确定"即可进行规范化，如图3-12所示。

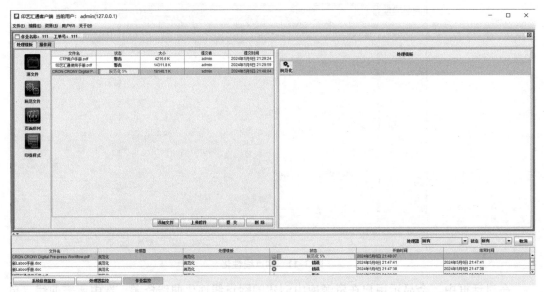

图3-12　规范化界面

提交成功后，文件开始进行规范化。可在窗口中的状态栏查看文件处理的进度，也可在作业监控里看到文件的详细处理进度。流程中提交的所有操作，也可以使用鼠标拖拽方式直接提交。操作步骤如下：首先选中要处理的文件，然后按住鼠标左键不放，直接拖动到右侧相应的模板上。这时，鼠标变成加号，松开鼠标，文件将会自动进行相应的处理。

"取消"操作可终止当前文件正在处理的操作。在源文件里，可取消文件的规范化处理。选中正在进行规范化的文件，右击选择"取消"选项，文件将停止规范化，在状态栏里将显示为"终止"。所有的取消操作均可在"作业监控"里执行。选中正在执行的某个处理任务，单击"取消"即可。

2. 规范文件窗口

在"规范文件"窗口中，显示源文件经规范化后，所生成的标准的单个PDF文件。在这里，可对文件进行预视，并可提交给自由拼版模板和输出模板。单击"规范文件"图标，界面将切换到规范文件的操作窗口，如图3-13所示。

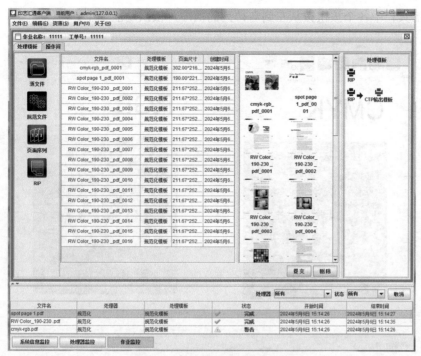

图 3–13　规范文件的操作窗口

窗口分为两部分，左侧以列表的形式显示了已规范化的单个 PDF 文件，可以查看文件的相关信息，包括文件名、处理模板、页面尺寸、创建时间。

（1）文件名：生成的标准的单个 PDF 文件的名称。文件名是通过源文件的名称、页数、类型来命名的，通过名称可以清楚地看出每个文件的页号及源文件类型。

（2）处理模板：源文件进行规范化所用模板的名称，可以查看该文件是由哪个模板处理生成的。

（3）页面尺寸：文件的页面大小，以宽 × 高的形式显示，单位为 mm。

（4）创建时间：文件的生成时间。

右侧显示文件的缩略图，每个文件都有一个相对应的缩略图。在左侧选中文件的同时，右侧的缩略图也同时被选中，缩略图可以预览该文件的大概内容。

在规范文件窗口中，可对文件进行预视、提交、取消、删除等操作。

3. 预视窗口

用户可通过文件缩略图粗略预视，也可通过打开预视对话框进行详细预视。

选中文件，单击鼠标右键，在弹出的对话框中选择"预视"选项或双击文件，弹出"预视"对话框，如图 3-14 所示。

在对话框中，可对图像进行放大、缩小、量尺、拖动、吸管、打印、镜像、反转等操作。

在对话框最上边是预视工具栏，可选择不同的工具进行预视；中间是文件的预览区域，显示文件预览图；最下边的状态栏中显示预视的操作信息，包括缩放倍数、使用量尺时的长度和角度、吸管显示的数值等信息。

预视工具栏中各个按钮按从左到右排列的次序，依次功能如下。

图 3-14 预视

(1) 放大按钮：鼠标单击 ⊕ "放大"按钮后，文件进行成倍放大，倍数最大为 16∶1，在对话框的状态栏中，可查看文件的缩放倍数。预视时，当光标为手掌时，按住 Ctrl 键，光标会变成放大镜，此时可以直接单击放大，也可在窗口中拉框选择区域放大，实现对图像进行局部预视（当使用"量尺"时，不能使用这种操作）。

(2) 缩小按钮：鼠标单击 ⊖ 按钮后，文件进行成倍缩小。倍数最小为 1∶16。同放大一样，预视时，当光标为手掌时，按住 Shift 键，光标会变成缩小状态，此时可以直接单击缩小预视。在任何状态下，都可直接按键盘上的"+""-"键对图像进行缩放预视。

(3) 量尺按钮：单击 按钮后，鼠标变为十字形状，这时可测量相应的对象。选中测量对象的起点，按住左键不放，同时拉动鼠标到测量对象的终点，在对话框的状态栏，将显示出该对象的长度和角度的具体数值。按住 Shift 键，可进行水平和垂直的直线测量。

(4) 拖动按钮：单击 按钮，光标转换为拖动状态，这时鼠标变为手掌，可拖动预览放大后的图像（预视对话框，打开时默认为拖动的状态）。

(5) 连拼按钮：预视文件连续拼接后的图像。这个操作只是针对进行连拼输出时进行预览。单击按钮后，将把该图像自动生成为三个竖向拼接在一起的图像，用户可查看图像拼接在一起后的效果是否正确。

(6) 旋转列表框：在旋转下拉列表框中，选择预视时合适的角度，可查看文件在旋转后的效果。

(7) 镜像复选框：选中复选框后，文件可进行镜像预览。

(8) 反转复选框：选中复选框后，文件前后和反转预览。"向前"和"向后"选项是仅在预视多个文件时起作用，单击后，可查看前一个文件或者后一个文件的预览图。"向前""向后"操作可使用键盘上的左右方向键来控制。

（9）文件显示框：在工具栏的最后，显示预视文件的名称，多个文件时，可在下拉列表中，查看、选择所要预视的文件。

4. 文件信息窗口

可通过文件信息窗口查看文件信息。右击选择"文件信息"选项或按住"Ctrl+I"键，弹出"文件信息"对话框，如图3-15所示。在"文件信息"对话框中，显示了文件名称、页面尺寸及分色信息。

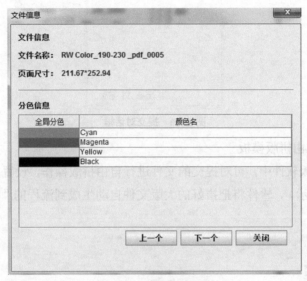

图 3-15　查看文件信息

显示文件的分色信息，包括全局分色、颜色名、延伸误差、专色转CMYK等参数。

（1）全局分色：以相应的颜色显示文件中的所有色版，包含C、M、Y、K四色及专色色版。

（2）颜色名：显示所有色版的名称。

（3）延伸误差：文件输出到输出器时，根据不同设备的需求对页面尺寸大小的调整。根据所设的数值使页面向左右延伸或缩短。

（4）专色转CMYK：主要针对有专色的文件起作用。对专色来说，选中后，文件中的专色转换为CMYK四色格式进行解释输出。这一操作需结合输出模板里"分色"参数来设置。

（5）更新所有页面：表示对话框中所设置的参数将对所有规范文件起作用。设置完成后，单击"确定"，即可完成文件信息的设置。

操作时，按住Ctrl键和Shift键可选择多个文件，按住Ctrl+A键选择所有文件，单击"删除"按钮或单击右键选择"删除"选项，系统将根据所选文件弹出相应的提示，单击"确定"后即可删除文件。删除文件时，正在使用的文件，不能被删除。文件删除后不能还原。

5. 提交规范化文件

在规范化窗口中,可将规范化好的标准 PDF 文件进行拼版和输出操作。这是流程的第二步提交操作。单击"提交"按钮或右击选择"提交"选项,打开"提交"对话框(如图 3-16 所示),在对话框中显示了所有可以提交的模板。

图 3-16 提交对话框

(1)提交给自由拼版模板

在佳萌自由拼版软件中,可对提交的文件进行自由拼版操作,设置好后,执行"拼版"操作(如图 3-17 所示),软件将把拼好的大版文件自动生成到流程的"印张样式"界面中。

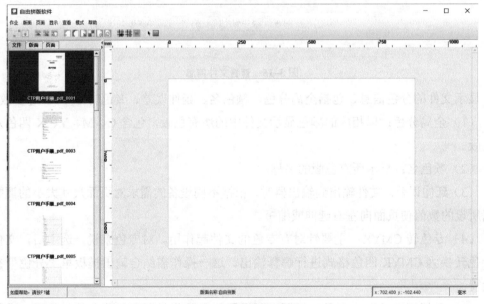

图 3-17 "拼版"操作

(2)提交给输出模板

流程中输出操作包括 PDF 输出、TIFF 输出、数码印刷输出、CTP 输出这四种。其中 PDF 输出与 TIFF 输出是将文件直接解释输出到一个网络共享文件夹中;而数码印刷输出与 CTP 输出拥有自己的输出器,然后流程将文件解释输出到输出器中。

当提交给 PDF 输出与 TIFF 输出模板时，模板中设置的路径必须可以正常连接且有文件创建权限，这样才能正确输出。

当提交给数码印刷输出和 CTP 输出模板时，首先要确定输出器是否已成功启动，当没有启动时，右边的图标将显示成灰色状态，提交后，系统会提示"输出器未打开，无法输出"，成功启动后，图标显示为深蓝色，这样才可进行输出操作。选中相应模板后，单击"确定"按钮，进行提交，提交后，在"作业监控"里可以看到解释输出的进度。

文件输出完成后，将自动生成到相应的共享文件夹下或相应的输出器中。

加网参数设定通过制版输出流程的输出模板来完成，其中科雷佳盟流程软件通过 CTP 输出模板的 RIP 设置来实现，如图 3-18 所示。

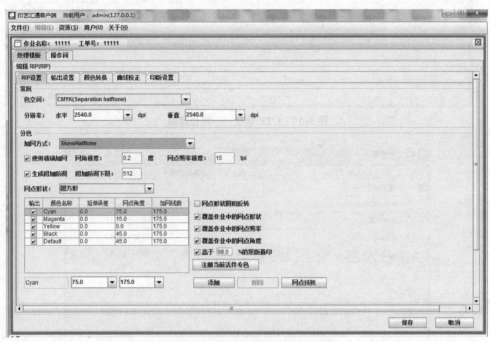

图 3-18　CTP 输出模板的 RIP 设置

网点形状在对应对话框中的"网点形状"下拉列表选择，网点角度与加网线数根据不同的色版在列表里进行修改。对于专色版可以通过"添加"按钮设置加网角度与线数。

二、工作流程转换颜色

流程中建有自己的色彩管理模块，通过各个设备的色彩特性描述文件 ICC Profile 实现不同颜色模式文件在不同的设备上的正确输出。在"颜色转换"设置页（见图 3-19）中有三个下拉列表选项，每个选项可用来选取一个颜色特性描述文件。

1. 源 RGB

选择被转换的颜色为 RGB 模式时，使用源 RGB 色空间特性文件。单击"浏览"按钮，弹出"ICC 文件选择"对话框，如图 3-20 所示。

图 3–19 CTP 输出——颜色转换

图 3–20 ICC 文件选择

在对话框中，列出了资源库中的具有 RGB 色空间 ICC 特性文件资源，选中所需 ICC 文件，单击"确定"即可。流程默认为"sRGB Color Space Profile"。

2. 源 CMYK

源 CMYK 是选择被转换颜色为 CMYK 模式时，使用的源色空间特性文件。选择文件方法与"源 RGB"项相同。

3. 设备色空间

选择要转换到的目的设备色空间特性描述文件。通常应该选择一个印刷机的特性文件（Profile）。选择文件方法同上。

色彩管理是用来在两个设备之间进行色彩转换，如果某颜色在 A 设备上的值为 Ca，要把它转换成在 B 设备上的值 Cb，就要借助 A、B 两个设备的彩色特性描述文件。其中 A 设备的彩色特性描述称为"源色空间"，B 设备的彩色特性描述称为"目的色空间"。换言之，要想校正一个印刷机的色彩，使之与某传统打样机色彩一致，则打样机为"源设备"，印刷机为"目的设备"。

4. 色彩转换意向

可根据不同的输出要求选择所需的色彩转换意向。默认为 Perceptual。

Perceptual：知觉法，从与设备无关的色空间向设备的色空间映射，将所有的颜色等比例的压缩，色域外的颜色映射到设备的色域范围内，此方式适合图像的颜色转换。

Absolute：绝对色度法，将保持色域内的颜色不变，并把色域外的颜色压缩到目标色域的边界上，此方式适合颜色的准确复制。

Relative：相对色度法，用于颜色准确和与介质相关的复制。将白场和黑场映射到目标色域，它可以改变亮度。

Saturation：饱和度法，增加色彩的纯度和饱和度，但色彩复制性差，此方式适合只注重颜色鲜艳的图形的复制。

【忽略内嵌 ICC】

该选项表示在解释输出过程中，忽略文件中内嵌的 ICC。为了避免因生成源文件时错误嵌入了 ICC 文件，而造成颜色不准或文字转为四色黑的问题，通常选中此项。

【保持纯色黑】

该选项是为了避免由 RGB 向 CMYK 转换时将黑色图形或文字对象转换成为四色黑而造成套印困难。

三、字库管理与设置

字库是流程中的一种重要资源，字库安装正确与否直接影响文件规范化后的正确性。流程在安装过程中，会自动安装标准字库，对于 PS 文件，用户如果在生成时自动下载了字体，则不需要再进行字体安装，而 PDF 文件有时并没有内嵌字体，所以要通过手动来进行安装，才能使文件解释准确。打开"资源库"后，单击"字库"标签页，窗口中以列表的形式显示所有已经安装的字体。

（1）安装字体：单击"安装"按钮，将弹出"安装字体"对话框，在窗口上方有四个基本选项。

源安装字库路径：设置准备安装字库文件的位置。"安装字库"的功能只用于装入 TrueType 字库，单击"打开"按钮，用它来浏览文件夹，帮助用户寻找字库文件路径。

TrueType 安装为：该选项说明 TrueType 格式字库按照什么格式装到流程中。有 Local Joinus Fonts by default（默认选项）、PostScript CID 2 font by default、PostScript TYPE 42 font by default 等选项。

流程安装字库时会将 TrueType 字库重新组织成新的字库格式装入。一般而言，如果要安装的是中、日、韩字体，可以选择安装为 CID 2 字库；如果要安装的是西文字体，可以选择 TYPE 42 字库，选择该选项后则无须选择 CID 编码。字体资源和安装字体如图 3-21、图 3-22 所示。

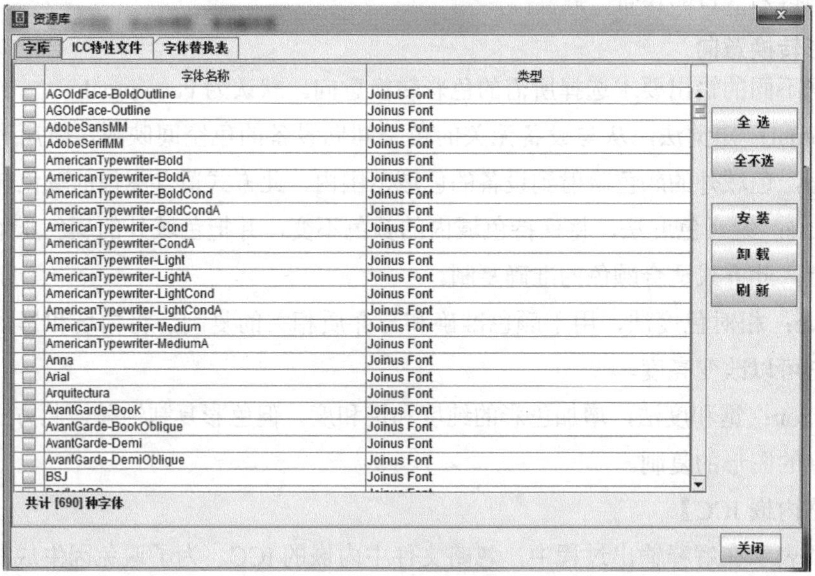

图 3-21　字体资源

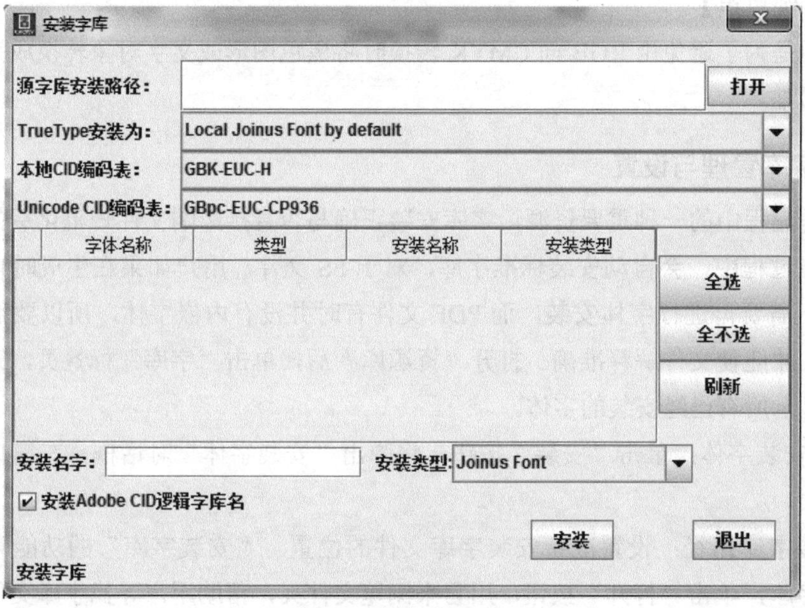

图 3-22　安装字体

为了更好地处理补字、竖排等问题，另外构造了一套独特的字库格式，称为"JoinUs Fonts"，大多数常用 TrueType 字库都可安装为该格式，因此"Local JoinUs Fonts by default"为默认选项。用户一般选择该选项即可正确安装字库。

本地 CID 编码表：该选项设置的是本地编码到 CID 字库的映射方式，本地编码指的是各地区区域性编码，对于中文而言有 GB2312、GBK、BIG5 等，日、韩也有各自的本地编码。针对不同的地区有很多种选项，但中国内地用户安装字库（无论是 GB2312 字库还是 GBK 字库）一般选用默认选项即可，默认为"GBK_EUC_H"。

Unicode CID 编码：由于 TrueType 字库一般按 Unicode 编码组织字库，需要设置 Unicode 到本地编码之间的映射方法。用户选择默认选项即可。默认为"GBpc_EUC_CP936"。

字体列表窗口：窗口中部是字体列表窗口，当从"源字库安装路径"中指定好字库位置后，系统开始搜寻该文件夹，将搜索到的字体依次列出，并在最下方显示出搜索到的字库文件个数，同时还列出这些安装字体的各种信息，包括"字体名称"，如 HYf0gj、"类型"、"安装名称"为安装到流程中之后的名称，"安装类型"即为"TrueType 安装为"中选择的类型。

当单击字体列表窗口中的一款字体时，字体列表窗口下部的"安装名字"和"安装类型"两栏中会出现该字库的相应信息，用户可在这两栏中自行更改安装名字和安装类型。但一般情况下，用户无须更改。

字体列表窗口中列出的每种字体的前面都有一个选择框，单击该选择框可决定是否安装该字体，列表右边的"全选""全不选"按钮是用来辅助选择要安装的字体，被选中的字体选择框内会打上"√"。"刷新"按钮用来刷新字体列表窗口中的内容。

窗口最下方还有一项。

注册 Adobe CID 逻辑字库名：表示在安装字库的同时，生成该字库的逻辑字库名。该功能是针对某些排版软件设置的，这些排版软件在处理字体时使用的是逻辑字库名，缺省选中。

以上内容都设置正确后，单击"安装"按钮，则选中字体就会被安装到流程中。安装完成后，在"资源库"的"字库"标签中将能看到正确安装的字体。

（2）卸载字体

在"字库"标签页的列表窗口中，通过前面的复选框选中要删除的字体，这时在字体前面的选择框内会打上"√"，也可单击列表右边的"全选""全不选"按钮来选择卸载的字体。选择完后，单击"卸载"按钮，此时会弹出警告框，确认无误后，单击"确定"按钮，则选中的字体将从流程的资源库中删除。

四、拼版

优联合折是优联数字化工作流程中的折手拼版版式设计软件，负责为流程中的折手拼版创建折手版式，包括纸张尺寸、版式布局、小版属性、模板标记、折页方式，等等。

1. 新建折手版式

在页面列表上单击右键，打开新建版式对话框，如图 3-23 所示。

图 3-23　新建折手版式

（1）"模板信息"设置框

模板名称：版式文件的名称。

装订方式：骑马订、胶订、自由订。

胶订：胶订可用于平装书之类的作业，装订时将各书帖平行叠加在一起。

骑马订：骑马订可用于小册子、纲要和目录之类的作业。装订时将各个书帖嵌套在一起的一种装订方式，如图 3-24 所示。

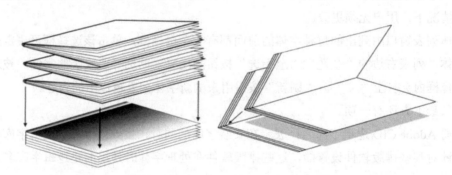

图 3-24　骑马订装订

自由订：该装订样式用于非折叠帖模板，如海报、拼合等拼帖工作。在自由订中，可以为拼帖作业在印张上组合成不同页面大小和方向的模板。

（2）"印张信息"设置框

纸张尺寸：实际印刷用纸的尺寸。

印刷方式：单面、双面、自翻、滚翻。

单面：对于单面印刷方式，印张只有正面印刷，常用于海报、名片、标签等。

双面：是最常见的印刷方式，用不同的印版印刷纸张的正反面。在版式正反面上，小版的正背位置镜像对称，如图 3-25 所示。

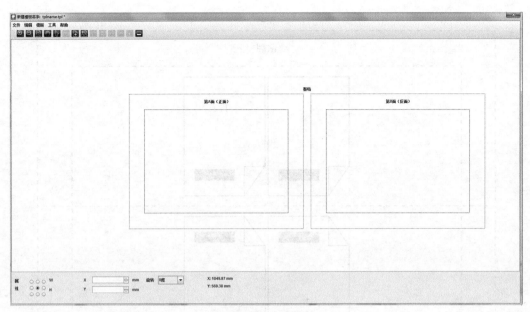

图 3-25 双面

自翻：对于自翻印刷方式，拼版的两面都位于同一个印版上。拼版的正反面相对于纸张垂直中心线左右对称，如图 3-26 所示。

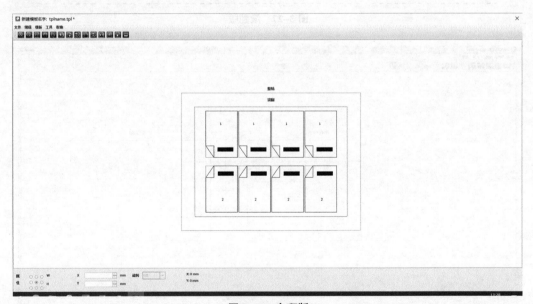

图 3-26 自翻版

滚翻：对于滚翻印刷方式，拼版的两面都位于同一个印版上。拼版的正反面相对于纸张水平中心线上下对称，如图 3-27 所示。

设置好模板及印张信息后，单击确定，可以打开如下拼版设计界面。

折手界面，如图 3-28 所示，共包括四部分：菜单栏、工具面板、设计面板、属性面板。设计面板中黑色虚线框为我们所设置的纸张尺寸。

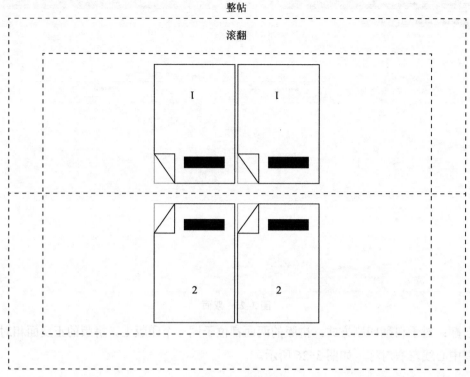

图 3–27 滚翻版

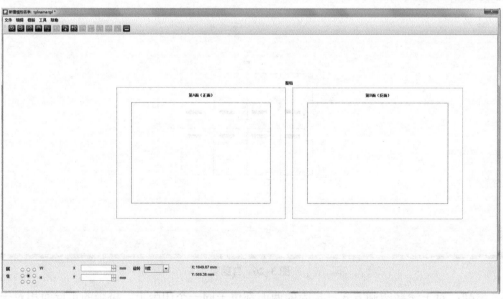

图 3–28 折手工作界面

2. 工具栏

工具栏共有 14 个按钮,如图 3-29 所示。按图中从左到右的排序,每个按钮的功能依次如下所示。

图 3-29　工具栏按钮

(1) 放大：放大拼版视图；

(2) 缩小：缩小拼版视图；

(3) 适合窗口：缩放帖视图以适合窗口；

(4) 拖动：移动窗口视图；

(5) 选择：用于选中页面或者标记；

(6) 设置页号：用来自动地依次添加页号；

(7) 添加单独页面：用来向版面内添加单独页面；

(8) 添加拼版：用来向版面内添加折手拼版；

(9) 页间距：控制显示和隐藏页间距显示框；

(10) 上下居中：使拼版上下居中于纸张；

(11) 左右居中：使拼版左右居中于纸张；

(12) 显示/隐藏页面：控制页面的显示和隐藏；

(13) 显示/隐藏标记：控制标记的显示和隐藏；

(14) 叼口位置：控制纸张的叼口位置。

3. 创建拼版

执行"模板"→"创建拼版"命令，打开创建拼版对话框，如图 3-30 所示。

(1) "成品尺寸"设置框

输入成品尺寸的宽度和高度。印张上的拼版页面大小都相同。无法更改一个页面的大小，但是可以更改拼版中所有页面的大小。

(2) "拼版布局"设置框

输入要添加到印张一面的水平和垂直拼版页面数。

书帖数量：用来控制一个印张上含多个书帖的拼版布局。

(3) "页面方向"设置框

单页方向：指定拼版页面的定向方式（相对于其他页面而言），这里是指左上方页面的方向。组中的剩余拼版页面将相对左上方页面来定向。

整体样式：用来设置整体页面的方向，包括头对头、脚对脚、头对脚。

4. 设置页号

拼版建立后，所有的页面的页号默认为1，用户需要根据折页方式设置页号，可以有两种方法。

(1) 手动设置页号

双击要修改的页面，会弹出设置页号对话框，可以同时设置小页正背面的页号。

图 3-30　创建拼版

（2）自动设置页号

利用设置页号工具自动添加小页页号。选择"设置页号"工具，单击要设置页号的小页，将自动为小页的正背面设置页号，页号随着单击依次递增。左键双击 ，弹出"设置页号"对话框可以设置起始的页号。

5. 设置页间距及页边距

对于创建的拼版，用户可以设置拼版的页间距和页边距，在工具面板上，单击"显示／隐藏页间距"图标。在拼好版的上边及左边出现小页间距数值。

左键单击页间距数值，打开"设置页间距"对话框（见图3-31），可以对页间距数值进行修改。修改页间距后，可使用左键单击工具面板中的"上下居中"和"左右居中"图标，可以使拼版上下和左右居于纸张中部。

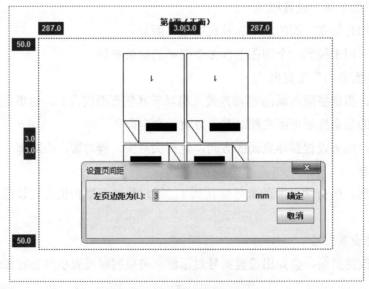

图 3-31　"设置页间距"对话框

6. 添加标记

为拼版添加裁切、套准、折叠、拉规、背帖、文本、重复以及自定义标记，设置选项如图3-32所示。

（1）裁切标记

"裁切类型"：可以为裁切标记选择单线和双线两种类型。

"长度"：裁切标记的长度，默认长度为6 mm。

"线宽"：裁切标记的线宽，默认宽度为0.25 pt（0.088 mm）。

"偏移出血值"：裁切线距出血位的距离。通常设置为0，即裁切线加在出血以外。

"正面""反面""正反面"指示标记将显示在印张上的哪一面。

图3-32 添加裁切标记

（2）套准标记

在对话框的左侧有套准标记，如图3-33所示。功能是为拼版页面添加套准标记，用来检查印刷机套印是否准确。

套准标记共有四种类型，分别是Nge-Cross（反十字线）、Circle Mark（圆形）、Hollow Circle（圆孔形）和Solid Circle（实心圆）。同时可为套准标记的宽度、高度和标记线条的宽度进行设置，即由"标记宽""标记高""线宽"命令来实现。而"偏移页面值"选项用来设置标记距页面的距离，默认值为3，为保证套准标记在出血以外，这里的偏移值一般为拼版的出血值。"正面""反面""正反面"选项用来指示标记将显示在印张的哪一面。套准标记通常加在正反面。

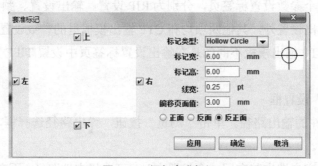

图3-33 添加套准标记

（3）折叠标记（如图3-34所示）

其参数如下所示。

"长度": 折叠标记的长度;

"线宽": 折叠标记线条的宽度;

"中心十字线": 位于折手拼版折叠的中心点, 用来检查折页是否准确, 这里可以选择是否添加十字线标记;

"标记颜色": 用来设置折叠标记的颜色值;

"所有分色": 标记将出现在所有分色版上, 可以在后面的文本框中设置颜色的色值;

"印刷色": 标记只出现在 CMYK 四色版上, 使用者可以分别设置标记 CMYK 分色的颜色值。

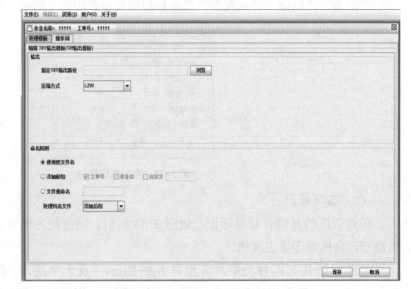

图 3-34　添加折叠标记　　　　图 3-35　编辑 TIFF 输出模板

五、导出为文件

1. TIFF 输出模板

TIFF 输出模板是将文件解释输出为 TIFF 格式文件, 对应的对话框如图 3-35 所示。在模板中, 分为五个参数设置标签页, 分别为 RIP 设置、输出设置、颜色转换、曲线校正和印版设置。除输出设置外, 其他四项均和 CTP 输出模板里的参数设置一致。

输出设置与 CTP 输出模板不同的是在输出设置标签页中设置 TIFF 文件输出路径及文件的命名规则。

（1）"输出"设置框

设置 TIFF 文件的输出路径。单击"浏览"按钮, 弹出路径选择框, 选择本地路径或网络共享路径即可。

当服务器端与客户端安装在同一个机器上时, 可选择本地路径, 不在同一台机器上时, 则需要选择共享路径。

（2）"命名规则"设置框

命名规则有三种：使用原文件名、添加后缀、文件重命名。

"使用原文件名"：输出的 TIFF 文件名直接按照流程中规范化或拼版后的文件名来命名。

"添加后缀"：在文件名后可分别加上工单号、作业名、自定义命名的后缀。后缀是在同名文件名后加 _1，依次排列。

"文件重命名"：对输出的 TIFF 文件设置一个自定义的名称。

"处理同名文件"：对于输出后的同名文件进行设置，可对文件进行添加后缀和覆盖操作。

2. PDF 文件输出

将流程中的规范文件或拼好的大版文件输出为标准的 PDF 文件。文件成功输出后将在指定的路径下建立一个与工单号同名的文件夹来存储这些 PDF 文件。图 3-36 所示为 PDF 文件输出设置对话框。

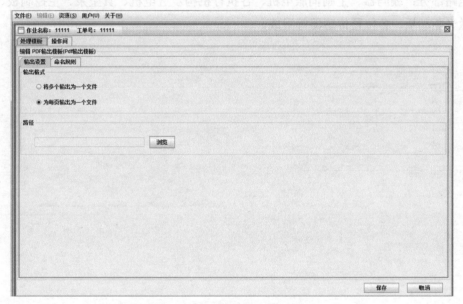

图 3–36　PDF 文件输出设置对话框

（1）"输出格式"设置框

用来设置文件以什么格式进行输出。"将多个输出为一个文件"选项是当选择多个文件同时输出时，则把这些文件合并为一个 PDF 文件；"为每页输出为一个文件"选项是当选择多个文件同时输出时，则把这些文件分别以单个文件的方式来输出。

（2）"路径"设置框

用来选择 PDF 文件的存储路径。

单击"浏览"按钮，选择网络共享文件夹即可。当服务器端与客户端安装在同一个机器上时，可选择本地路径。"单页文件"设置框中的选项与 TIFF 输出设置相同。

第五节　制版机及外接设备

一、制版机的基本构造

以科雷的直接制版机热敏系列TP-46、TP-36、TP-26系列为例，CTP直接制版机的构造基本由光学部分、机械部分与电路部分三部分构成，如图3-37所示。

光学部分：主要包括激光器、光纤耦合、密排头、光学镜头、光能量测量等。

机械部分：机架、光鼓、墙板、送版部、版头开闭部、装版辊部、卸版部、排版部、丝杠导轨部、扫描平台部等组成。

电路部分：编码器、主副伺服电机、各执行机构步进电机、真空泵、主控制板、接线板、激光驱动板、各位置传感器等组成。

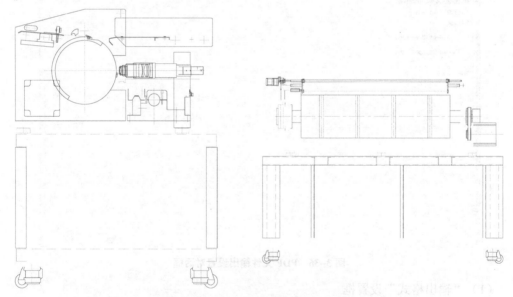

图3-37　CTP直接制版机的结构示意

通过USB通信线将CTP制版机连接至计算机，连接线位于罩壳右下侧USB接线端口，采用数据接口螺钉固定，确保连接可靠。

二、供版机

CTP供版机是实现纸版分离、自动进版的设备，用来实现CTP制版设备的自动供版操作，其结构如图3-38所示。

1. 供版机的连接

（1）连接供板机通信线，位于主机 USB 接线端口侧；

（2）连接供板机电源线，位于主机 USB 接线端口侧，电源备用端口。

图 3-38　供版机结构

2. 供板机工作模式设定

（1）连接供版机设备需要在工作模板中勾选供版机选项，可以进行供版机工作设置；

（2）设置单版盒供版机参数：设置供版盒吹纸吸版压力，根据版材大小及纸张重量选择合适的工作压力，有高、中、低三档可选；

（3）设置多版盒供版机参数：设置供版盒吹纸吸版压力，根据版材大小及纸张重量选择合适的工作压力，有高、中、低三档可选；设置供版机版盒参数，最多支持 5 版盒。

三、内置打孔

内置打孔为选配结构，是一种可以集成安装 CTP 设备内部，完成同步套印打孔的模块。内置打孔的结构如图 3-39 所示。

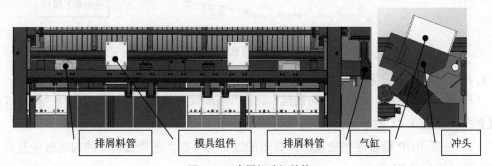

图 3-39　内置打孔机结构

1. 内置打孔的连接

（1）单击软件操作界面的"辅助设置"项，调出"外部设备设置"对话框，选择"打孔过桥"设置界面；

（2）选择对话框中的"内置打孔"，在"选择过桥"的勾选框中打钩；

（3）无须设置过桥类型及左侧过桥参数。

2. 内置打孔工作设置

（1）包含内置打孔功能的设备需要在工作模板中勾选"打孔过桥"选项，可以进行内置打孔工作设置；

（2）设备编号：选择"内置打孔"；

（3）进口方向、出口方向：内置打孔无须设定；

（4）模具选择：用来选择当前工作模具的组号，仅支持单组工作，最多支持 5 组模具安装。

3. 内置打孔的维护

（1）定期观察铝屑是否有毛边，必要时修模，延长模具使用寿命；

（2）定期检查气缸滑杆润滑及密封情况；

（3）每 24 个月更换气缸密封件。

四、打孔过桥

打孔过桥为选配外设，是一种可以根据 CTP 设备设定，自动完成套印打孔的机器，主要功能有：自动匹配 CTP 制版设备使用的规格版材、自动套印打孔、四向版材输入/输出，并且自动连接到冲版设备。BGP 打孔过桥的结构如图 3-40 所示。

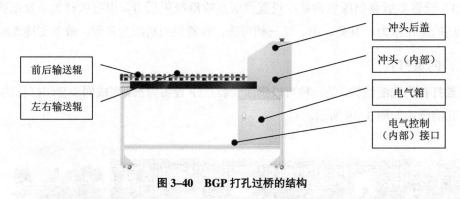

图 3-40　BGP 打孔过桥的结构

1. 打孔过桥辅助设置

（1）单击软件操作界面"辅助设置"调出"外部设备设置"对话框，选择"打孔过桥"设置界面（如图 3-41 所示）；

（2）依照 BGP 设备连接数量选择有效的过桥号（在"选择过桥"的勾选框中打钩），最多支持 4 组过桥连接；

（3）根据机型设置正确的"过桥类型"。

2. 打孔过桥工作设置

（1）连接 BGP 打孔过桥的设备需要在工作模版中勾选打孔过桥选项，可以进行打孔过桥工作设置，如图 3-42 所示；

（2）设备编号：用来选择当前 BGP 设备的顺序编号，本设备支持多组 BGP 设备联线使用；

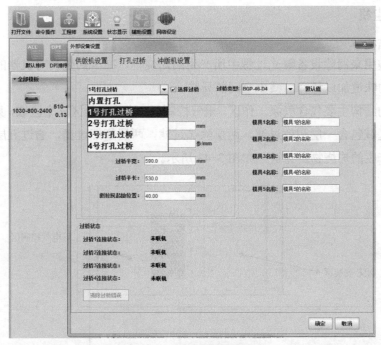

图 3–41 打孔过桥辅助设置

（3）进口方向：设定版材进入 BGP 设备的方向，本设备支持四向可选；

（4）出口方向：设定版材离开 BGP 设备的方向，本设备支持四向可选；

（5）模具选择：用来选择 BGP 设备当前工作模具的组号，仅支持单组工作，本设备最多支持 3 组模具安装；支起短过桥，后端位置与冲版机入口对齐。

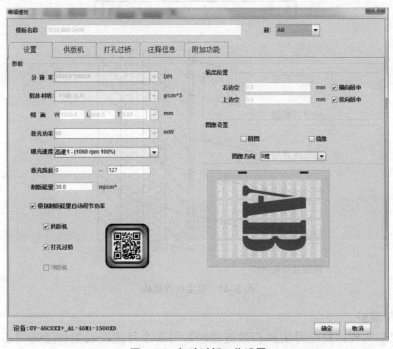

图 3–42 打孔过桥工作设置

五、吸尘机

1. 产品结构

吸尘机被用来清除设备曝光过程中所产生的灼烧物，减少灼烧物对机器的污染，同时对烧灼产生的味道加以稀释，减少空间的异味。

吸尘机的结构主要包含气泵工作区、电气控制区、滤芯安装区三个区域，见图3-43（上图）。滤芯区域包含一层初效过滤，两层除味过滤，两层超细过滤，通过五层过滤有效减少设备曝光产生的灼烧物及味道，如图3-43所示。

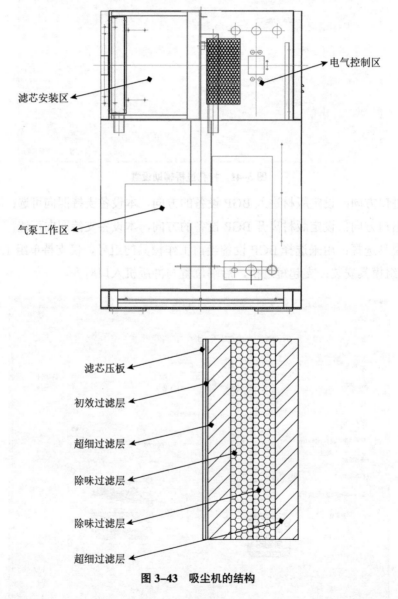

图 3-43 吸尘机的结构

2. 使用与维护

（1）定期清理：每个月更换初效过滤层，每3个月清洗吸尘和除味滤芯，每半年更换一次吸尘滤芯/除味滤芯。

①如设备使用环境或版材灰尘较多，滤芯保养及更换周期需适当缩短。

②如滤芯保养不当，可能会造成吸尘和除味效果下降，严重时影响CTP出版除尘效果，甚至带来CTP激光系统的损坏。

（2）设备工作温度较高，请保持设备环境温度，避免通风散热区围堵，影响设备正常运行，甚至造成设备损坏。

（3）保证吸尘管路通畅，每半年检测管路是否有老化/破损/扭曲等现象，必要时需及时更换。

3. 滤芯更换方法

（1）打开滤芯安装区盖；

（2）依次取出滤芯压板—初效过滤层—超细滤芯—除味滤芯，进行清洗或更换（清洗时使用干燥气体清洁，请勿用水）；

（3）放入清洗后（新）的滤芯；

（4）锁闭滤芯安装区盖，扣上锁扣。

4. 故障排查

产品在使用过程中如发现异常，请参见表3-1。

表3-1 故障排查及处理

故障现象	可能原因	处理方法
机器不工作	1. 电源没有开启 2. 控制线未连接或接触不良 3. 上位机控制选项未勾选	1. 开启设备电源 2. 检查或确认控制线连接是否良好 3. 正确设置上位机模板参数
吸力变小	1. 过滤器堵塞 2. 吸尘管连接漏气 3. 吸尘管老化破损	1. 清洁或更换过滤芯 2. 紧固抱箍，避免漏气 3. 更换破损老化的吸尘管
故障指示灯亮	1. 变频器报错 2. 变频控制器故障 3. 电机故障	1. 根据变频器报错信息进行问题排查 2. 更换变频器 3. 更换电机

第六节 冲版机

CTP印版输出后需要通过冲版机进行显影，显影成像的CTP印版才能进行上机印刷使用，因此在印版输出前需要调整与确认冲版机的状态。为达到最佳性能，冲版机及周围区域必须保持干净。应在每个工作日的开始和结束时清洁冲版机。

清洁时，请遵循以下规则：

（1）禁止使用硬毛刷、研磨材料、溶剂、酸性或普通碱性溶液来清洁辊轴和其他组件；

（2）只能使用规定的化学品来清洁冲版机；

（3）禁止使用砂纸、百洁布或其他研磨材料去除冲版机零部件上的污渍或物质；

（4）使用白色的无绒布来清洁冲版机。

一、检查设备

每周至少对冲版机所有部件进行一次仔细的检查。需要注意的是，在打开冲版机罩盖前，需先关闭主电源。表3-2显示了在检查冲版机和执行相关维护任务时要查看的内容。

表3-2 检查冲版机和执行相关维护任务的内容

故障部位	故障现象	处理方法
显影部位	显影槽中有严重的结晶残留	排干显影槽，用显影清洗剂浸泡冲洗
	显影液液位较低	增加补充液
	循环管不循环	检查喷淋管是否堵塞并做清理；检查循环泵是否正常工作
水洗部位	喷淋管堵塞	清理喷淋孔
	水量较小	加大进水开关阀门，或者增加抽水泵
	水质不干净	增加净水器
上胶部位	上胶管堵塞	清洁上胶管
	胶水杂质较多	更换新胶水
	胶水浓度更高	加水，增加稀释比例
烘干部位	印版从冲版机出来后，没有完全干燥	增加烘干温度
滚轴部位	运转有异常	如果是冲版机内滚轴异响，可以调节松紧度或者添加凡士林；如果是冲版机外滚轴异响，可以添加机械黄油
冷凝部位	显影液降温不良	检查冷凝液液位，并添加冷凝液；如冷凝液杂质较多，更换冷凝液

二、显影部分维护

需要注意的是，显影剂有腐蚀性。需戴上安全镜、丁腈手套，穿防护服。根据当地相关法规处理废弃的显影药液。首先排干显影液药槽；其次清洗显影部分。当执行维护和服务时需清洗显影部分。清洗显影槽前，必须确保显影液已排尽，排液阀被打开。在放入排液槽或空的容器里时排水管必须是空的。最后，更换显影过滤器（如图3-44所示）。需要注意的是，在执行显影部分操作时需穿戴相应的防护衣服、眼镜和手套。

1-显影过滤器出口阀（打开）；2-显影过滤器进口阀（打开）；3-压紧—释放按钮

图 3-44　显影过滤器

三、水洗部分维护

水洗部分的维护首先排干水洗药槽；其次清洗水洗部分。在执行周期性的维护和服务时，应清洗水洗部分。最后更换水洗过滤器。

四、上胶部分维护

上胶部分的维护，要防止胶料变硬，需在每个工作日结束时，执行以下操作：①将剩余胶料放回容器；②冲洗上胶部分。

五、冷凝器部分

冷凝器位于冲版机底架前面，用来冷却显影部分中的显影液。冷凝器（见图 3-45）由制冷单元（压缩机、散热器和风机）、冷凝器相关组件组成。

1-压缩机；2-散热器；3-风机；4-泵；5-排液管；6-排液阀

图 3-45　冷凝器

第四章 印版的输出

1. 能按工艺单设定输出参数；
2. 能对栅格图像处理器（RIP）处理后的数字文件进行检查；
3. 能完成印版的显影；
4. 掌握印版输出过程中出现的问题的解决方法。

第一节 CTP 印版输出

CTP 制版机开机步骤如下：①要打开 UPS 电源，观察温/湿度计指针是否处于绿色的位置；②打开 CTP 右侧红色电源开关；③打开与制版机连接的计算机电源；④打开 CTP 制版机上面的轻触式开关，等待机器自检 5 分钟；⑤打开驱动的软件，检查软件和机器的状态，如图 4-1 所示。

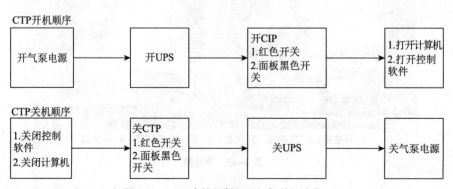

图 4-1 CTP 直接制版机开机与关机流程

一、启动

CTP 制版机使用时,首先打开 UPS 电源,之后打开冷气。接着打开空气压缩机,观看指针是否处于绿色的位置;开启设备电源使用凸轮开关的设备,将凸轮开关打开到如图 4-2 所示位置。

使用微型断路器的设备,将断路器向上打开到导通位置,接入设备电源(右侧为 3 相电源断路器),如图 4-3 所示。

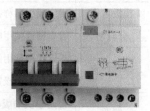

图 4-2　凸轮开关　　　　　　　　　　图 4-3　电源断路

打开 CTP 上面的轻触式开关,等待机器自检 5 分钟。在"计算机管路"→"设备管理器"展开目录下,确认制版机设备驱动程序已经正确安装,正确安装时,在通用串行总线控制器下可以找到以下硬件"Cron Laser device,V1C",如图 4-4 所示。

如果安装不正确,软件将无法连接设备,这时可以通过收到更新驱动程序的方式来人为协助安装制版机设备驱动程序,软件驱动程序文件存放于软件根目录下-"USBDrv128B"文件夹内,请手动更新,直至设备驱动正确安装。

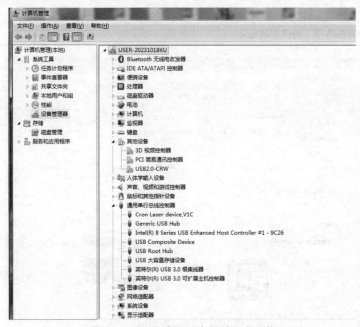

图 4-4　确认制版机设备驱动程序安装

正确安装完软件后,双击桌面的 Laboo5.1.0 软件快捷方式图标,弹出启动界面,加载完成启动程序后即完成启动 Laboo 输出器。

二、设定初始模板

新装软件首先需要设定初始模板，单击新建模板对话框，设定一个初始模板：根据机型规格设定允许范围内的模板，确认输入"模板名称""分辨率""载体材质""幅面""制版能量""曝光速度"等参数（如图4-5所示），参数说明见表4-1。

表4-1 设置参数

名称	说明
模板名称	设定模板名称；建议规范命名：版材幅面—分辨率—版材厚度。如：510-400-2400-0.135
分辨率	设定模板使用的图像分辨率（机型配置不同，支持的分辨率也不同）
载体材质	设定模板使用的版材材质，支持材质：PS版、胶片（机型配置不同，支持的材质也不同）
	注意：材质设定错误会造成离焦、设备抖动等问题，影响成像质量
幅面	设定模板的版材幅面尺寸，尺寸方向参考右下示意图，厚度依照实际测量为准（最大最小幅面由机型决定）
激光功率	激光功率为手动设定值，依照版材的能量需求而定，其值不超过设备出厂功率
	注意：如勾选根据制版能量自动调节则无须手动设定，建议选择
曝光速度	曝光时的光鼓转速选择，依照实际生产设定，其最高速由制版能量、最大功率及最高限速决定
激光路数	设定路数0～128（激光路数由机型决定）
制版能量	制版能量由版材特性决定，由版材供应商提供或实测而得
供版机	自动供版选项，如勾选，则该模版工作时包含供版机自动供版（由机型决定）
打孔过桥	打孔选项，如勾选，则该模板工作时包含自动打孔功能（由机型决定）
冲版机	冲版机选项（由机型决定）

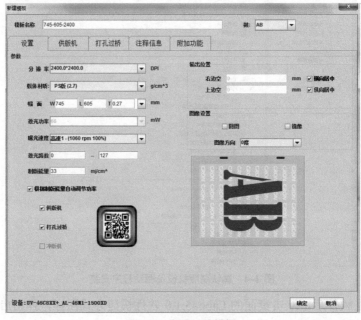

图4-5 新建工作模板

新建工作模板：单击"新建模板"按钮，调出新建模板对话框，根据实际的工作需求设定一个工作模板；设定完成后确认，建立一个热文件夹可以便于后续的批量操作。

三、编辑工作模板添加作业

编辑工作模板：工作模板设定完成后，如需要更改模板参数，可以通过编辑模板功能实现模板的再编辑，编辑模板时，模板名称、载体材质、分辨率参数将会被锁定无法编辑，仅限于可调参数的更改。

通过添加 TIFF 文件添加任务：单击"打开文件"按钮，选择相应目录下的 *.Tiff 和 *.PPG 文件作业，选择需要使用的文件及下方对应的模板（需保证参数一致）；确认后，该文件即出现在软件的输出列表中，如图 4-6 所示。

图 4-6　添加文件

四、出版前检查

认真核对文件；根据工单选择正确的文件输出 CTP 版材；对 TIFF 文件进行最后的常规检查（角线、色标、尺寸、分色、叼口、针位线）。

在流程软件里设定版材输出参数时，应根据输出版材的实际需求设定出版功率、转速、图像位置等相关参数。

五、CTP 上版操作

依照软件操作说明，选择正确的模板，执行"模板调整"，将设备模板调整到对应位置；检查版材是否平整，有没有黑点白点；看清版材尺寸和计算机的提示操作；上版时，需要把版材的下沿对齐机器的边缘，然后按动进版键；下版时需要轻拿轻放，平整放到冲机上。

放置版材的步骤如下：

①将版材放置在简易供版台对应位置上，版材右侧边靠近居中靠规位置，离靠规约1mm；

②将一批次数量的版材（不超过版盒限制数量）放置到设备的供版台上，前端靠齐前规，右侧靠近居中靠规位置，离靠规约1mm。

六、出版操作

1. 设定输出参数

对系统进行参数设置如图4-7所示。

图4-7 参数设置

（1）单位设置即输出图像的计算单位，可选择毫米或英寸。

（2）保留信息行数：在设备执行动作时会自动保留任何返回的信息，本功能可设置需要保留的行数。在返回信息界面单击右键，可以清除所有保留信息。此处显示为保留信息1000行。

（3）保留文件个数：此处显示为保留文件200个。

（4）慢转光鼓启动时间：当设备处于待机状态下，启动光鼓慢转开始时间。

（5）慢转光鼓停止时间：当光鼓处于慢转状态下，停止转动的时间。

（6）热文件夹扫描间隔：系统自动搜索热文件夹时间周期，默认值为10秒。

（7）热文件夹默认路径：可设置热文件的默认状态指定哪个文件夹。

（8）信息栏双击弹出快捷菜单：选中此选项后在信息栏对话框内可右击选择操作。

（9）最后一个作业不自动供版：最后一个作业完成后，自动供版机不自动供版。

（10）自动供版输出时幅面切换提醒：不同幅面切换时，软件给予提醒。

（11）允许大版小幅面输出：选择本功能可以使用比图样尺寸大的版来输出该图样。

（12）大版小幅面输出确认：选择该功能后，大版小幅面输出前将提示用户，是否确认使用大版输出小幅面功能。

（13）错误状态下声音警告：如输出失败或出现错误，会不间断地发出警告声，移动鼠标，声音停止。

（14）是否每张激光检查：选择该功能后，在每张作业输出时检查激光功率。

（15）语言选择：界面语言有中文与英文。

（16）设置主机 ID：设置当前设备编号。

（17）网点补偿：对制版网点进行线性化补偿。

2. 辅助设置

（1）供版机设置

设置供版机的相关使用参数及状态显示如图 4-8 所示。

多用版盒：多用版盒是支持不同版材尺寸的多功能版盒；

1 号、2 号、3 号版盒：定义相关版盒使用的版材尺寸；

吸纸强、中、弱：根据版材衬纸的类型不同，可选择三档不同压力的吸纸方式；

吹气强、中、弱：根据版材衬纸的类型不同，可选择三档不同压力的吹纸方式；

供版机状态：当前供版机的联机状态。

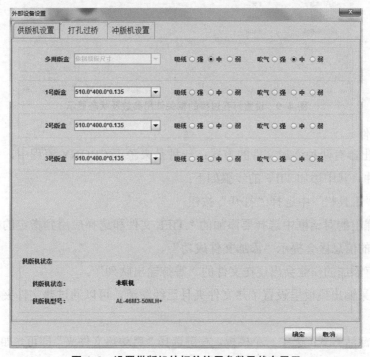

图 4-8　设置供版机的相关使用参数及状态显示

（2）打孔过桥（如图4-9所示）

①选择过桥：根据作业需要选择过桥；

②侧规最大版宽：当前过桥工作最大版材幅面；

③侧拉规电机步距：侧拉规步进电机1毫米所运动的脉冲数；

④过桥半宽：当前过桥支持最大版材幅面宽度的一半；

⑤过桥半长：当前过桥支持最大版材幅面长度的一半；

⑥侧拉规起始位置：工作状态下，侧拉规拉版的起始位置；

⑦模具1、2、3、4、5名称：自定义当前过桥模具的名称；

⑧过桥状态：当前过桥联机状态。

图4-9　设置打孔过桥的相关使用参数及状态显示

3. 添加文件

Laboo输出器有两种添加文件的方式：一种是来源于CRONY流程中，另一种是直接添加TIFF文件。其中添加TIFF的步骤如下：

（1）在"工具栏"中选择"打开"按钮；

（2）在弹出的对话框中选择要添加的*.TIFF文件和选择应用到指定的模板；添加成功后软件界面的信息区会提示"添加文件成功"。

（3）所有添加的作业会出现在文件的"等待输出队列"。

注意：如果输出模板里设置了热文件夹且已经激活，可以通过热文件夹向输出器添加文件。

热文件夹激活方式：选中一个模板，右键选择"启动热文件夹"即可，如图4-10所示。启动完成后模板左上方显示"HOT"文字。

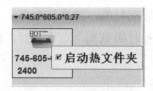

图 4-10 启动热文件夹

4. 输出作业

设备在输出文件之前会自动检查当前设备参数并调整为当前作业模板。

鼠标单击"单个输出",也可单击"连续输出",相同模板下的作业文件即从上至下按次序输出,已经完成输出作业将自动移到"已完成作业队列"中,所有操作过程信息将在工作信息栏中体现。当文件处于输出状态时,"单个输出"会变为"停止输出",单击"停止输出",设备停止输出工作。

补充实用操作:上版后如需取消制版,可在"准备曝光"时或曝光处理初始阶段按"停止"键。按下停止键后,设备不会自动卸版,需要在"用户操作"中单击命令卸版,使印版从设备中卸出。

光量偏移等报错提示,只需按确认键即可,某些报错按照提示进行相应操作即可。

5. 注意事项

(1) 将 TIFF 文档复制到相应尺寸的出版文件夹,放印版前先确认机器显示尺寸是否对应。

(2) 从箱内取版前应确保双手干净无灰尘、手面无水、干燥。

(3) 取版时,印版与衬纸一并拿出,否则版面易受到摩擦而产生白线。

(4) 取版放版时手指勿用力过大,且应轻拿轻放,防止折版。注意版面不要受到外物碰擦。

(5) 版放置上版平台前,应检查版面是否平整干净,尤其注意版边缘是否折翘,保证平整,否则极易造成制版机卡版。

(6) 放版时对好正确的位置尺寸,注意不要放斜,过于斜会导致上版过程中卡版。

(7) 盖门时应轻力,不可用力过大重扣,否则易导致激光器振动,版面上会留下长的蓝线。

(8) 上等候版应在前一张版 100% 曝光前完成上版操作。

(9) 出不同尺寸印版衔接时,按上版键前应调整至相应尺寸出版模式,否则会卡版,导致调节传感器零件扭曲变形报错,开闭门不能打开。

(10) 置顶某文件前,需先确认机器内已上的等候版是否与该文件的尺寸相对应。总之要保证等候版与下一个出版文件尺寸相对应。

(11) 推入冲版机时注意不要倾斜,否则会导致冲版过程中卡版,应及时按"急停"。

(12) 稍微使些推力,以保证前端胶辊充分咬住印版,不然印版会在前方平台停滞。

(13) 及时收版,以避免前后两张版叠版而刮伤版面。

(14)收版后,版与版之间应夹衬纸,防磨伤刮坏。

(15)原版和成品版都不可长时间见光,出完版后,将箱盖再盖回去。

6. 日常工作及维护保养

(1)出版需登记入账。

(2)每天早上核对前一天账单,补版单也需核对。

(3)每天早晚出版各测一次版。偏差范围应控制在 ±1% 之内,如网点偏差较大,原因可能是近期出版量过多导致药水药性提前减弱,或药水长时间没更换,或者是水槽外循环排水口被堵住导致水流入显影槽而造成药水药性减弱。采取措施如下:更换新药水,水槽外循环排水口若被堵应拆掉毛刷等构件后疏通排水口。排除药水和机器原因外,仍有误差则可能是文件本身有问题。

(4)保证地面干净。

(5)室内温度要控制好,保持在 21～25℃。

七、CTP 关机操作

使用计算机里的设备驱动软件关闭 CTP 制版设备;关闭 CTP 制版机的电源、计算机和切断电源。

制版机使用注意事项如下:

(1)CTP 制版机工作时不可关机,否则会卡版;

(2)放置 CTP 版时须将药膜朝上平行放入冲版口,不可放斜或反放。

第二节 设置印版的分辨率与加网参数

一、直接制版流程中印版输出分辨率的设置

直接制版流程中印版输出分辨率的设置通过输出流程模板的参数设置完成,如科雷佳盟流程软件就是通过规范化模版参数设置实现。

规范化模板是将源文件解释为标准的单个 PDF 文件。在模板中,分为四个参数设置标签页:常规设置、字体设置、图像压缩、PS(EPS)设置。其中的常规设置选项如图 4-11 所示。

1. 页面设置框

设置规范化页面的规格。该设定仅适用没有页面描述的 EPS、PS、PDF 文件,对于包含这些信息的文件不起任何作用。

（1）默认尺寸

在下拉菜单中，可选择相应的尺寸，如 A3、A4。也可自定义页面尺寸，选择"自定义"选项，在下边"长"和"宽"中分别输入相应的尺寸，以 mm 为单位。

（2）分辨率

该参数定义了标准化分辨率。如果将要处理的 PostScript 编码中没有包含任何的分辨率信息，就会使用标准分辨率。分辨率主要定义了 PostScript 阴影的加网线数。这对图像和字体没有任何影响。

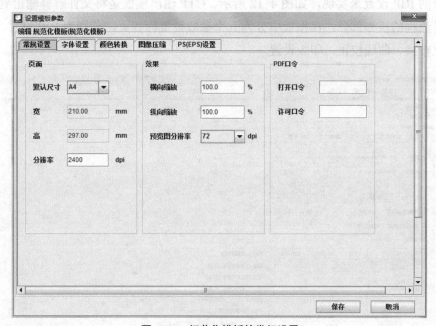

图 4-11 规范化模板的常规设置

2. 效果设置框

设定规范化页面的缩放倍率和生成预览图的分辨率。

（1）横向缩放

设置页面横向缩放的倍率。规范化时按照设定的比例，将文件进行放大或缩小。默认值为 100%。

（2）纵向缩放

设置页面纵向缩放的倍率，默认值为 100%。

（3）预览图分辨率

设置规范化页面后，生成预览图的分辨率。默认为 72dpi。可在下拉菜单中选择分辨率，如 72 dpi、144 dpi 等。

3. PDF 口令设置框

仅针对带口令的 PDF 文件。有些 PDF 文件中设置了口令，没有口令不能正常处理文件，所以必须在此处设定所需的口令，才能正常进行规范化。

(1)打开口令

输入 PDF 文件的打开口令。

(2)许可口令

输入 PDF 文件的许可口令。

二、直接制版流程加网参数的设定

加网参数设定通过制版输出流程的输出模板来完成,其中科雷佳盟流程软件通过 CTP 输出模板的 RIP 设置来实现,如图 4-12 所示。CTP 输出模板是将文件解释输出到 CTP 输出器,是流程的核心部分。在模板中分为五个参数设置标签页,分别为 RIP 设置、输出设置、颜色转换、曲线校正、印版设置。

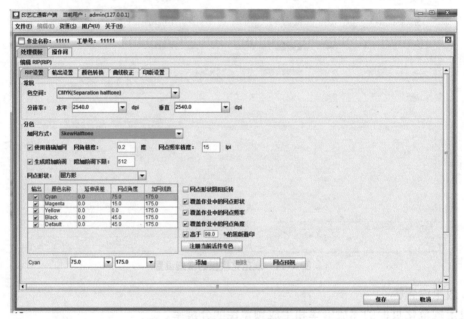

图 4-12 RIP 设置—实现加网设定

1. 常规设置项

(1)色空间

"色空间"反映了输出设备对色彩及图像数据的处理方式,其实质是告诉用户以怎样的方式处理待输出的文件:是否需要 RIP 分色,是输出二值加网数据还是连续调数据,等等。

在"色空间"下拉列表中可选择不同的色空间。例如,对于照排机、直接制版(CTP)机等单色设备只能选择 Gray(Halftone)和 CMYK Separations(Halftone);对于彩色喷墨打印机可选 CMYK(Halftone);对于 TIFF 输出所有的项目都可选用。

① Gray(Halftone)

表示单色加网模式,制版软件前端分色,无须 RIP 分色,生成二值点阵数据。

② CMYK Separations(Halftone)

表示 RIP 分色加网模式,前端制版软件生成的为复合色文件,需要 RIP 对其进行分色加网，为每个色版生成一个二值点阵图像。

关于"前端分色模式"和"RIP 分色模式"的说明如下。

对于 RIP 而言，所有照排机及直接制版（CTP）机都是单色网目调设备，因此当设备为照排机及直接制版机时，色空间只有以下两项可选。

Gray（Halftone）通常也被称为"前端分色模式"，要求排版软件前端分色，换言之，在排版软件中需要选择"分色"，生成的 PS 文件中已经包含了四个单色页面（假定只有 CMYK 四色），四个色版中的颜色数据都已确定，色版之间的压印（Overprint）、陷印（Trapping）等效果都已实现完毕，RIP 对其只能进行单纯解释，已经不可能影响其颜色、压印、陷印等效果。

CMYK Separations（Halftone）通常也被称为"RIP 分色模式"。要求排版软件生成复合色 PS 文件，即 PS 文件中包含单个彩色页面，对于这类文件，RIP 可以有很大的作用空间，可以对其进行色彩管理；对专色进行灵活处理；由 RIP 控制生成压印、陷印效果，等等。PostScript 3 标准中的很多新增指令都要基于复合色 PS 文件完成，并且在印前作业流程中会发挥更大的作用。

③ Gray（8 bit contone）

单色连续调模式，解释成灰度图像。

④ CMYK（Separations 8 bit contone）

RIP 分色连续调模式，前端排版软件生成的为复合色文件，需要 RIP 分色，为每个色版生成一个灰度图像。

⑤ CMYK（8 bit contone）

RIP 分色连续调模式，前端排版软件生成的为复合色文件，需要 RIP 分色，生成一个彩色图像。

⑥ RGB（8 bit contone）

输出 RGB 彩色图像模式，与上一项类似，前端排版软件生成的为复合色文件，需要 RIP 转换为 RGB 模式，并为每个色版生成一个彩色图像。

（2）分辨率

设置输出文件的分辨率，由后端输出设备决定。从下拉菜单中选择合适的水平和垂直分辨率，也可自定义输出分辨率。RIP 将按照选定的分辨率来生成点阵文件。当选择的分辨率与输出设备实际分辨率不一致时会造成输出尺寸变形。

2. 分色设置项

该设置项的参数是否可使用，取决于"色空间"中设置的分色方式。在某些模式下，分色参数是无效的。

（1）加网方式

在下拉列表中，列出了"Skew Halftone""HQS Halftone""Balance Halftone""External Halftone""Normal Halftone""LessIntaglio Halftone" 6 种调幅加网方式。

不同的方式对加网精度、网点玫瑰斑以及灰度层次的控制算法不同，"Skew Halftone""HQS Halftone""Balance Halftone"为最常用的加网方式。"External Halftone"为自定义网点加网方式，多用于凹版制版的加网。

（2）使用精确加网

选中此项后，可对"网角精度"和"网点频率精度"进行设置。

①网角精度是指 RIP 加网后实际网点角度与设定网点角度之间的允许角度差。系统默认值为 0.2°。

②网点频率精度是指 RIP 加网后实际网点频率与设定网点频率之间的允许频率差。系统默认值为每英寸 15 线。

以上两项设定用户不得自行更改，否则可能引起撞网现象。

（3）生成附加阶调

选中此项后，可对"附加阶调下限"进行设置。缺省值为 512。输入的是灰度级数下限，灰度级数越高，输出图像的层次尤其是渐变的平滑程度越好，但对解释速度和网点形状有一定影响。

（4）网点形状

在下拉列表中，列出了 31 种不同的网点形状，可根据需要进行选择，还可通过单击模板下方的"网点预视"按钮，预览在不同网点角度、不同加网线数下的网点形状。

当加网方式为 External Halftone 时，网点形状将载入在资源库中存储的"自定义网点"资源，可选择其中一个进行加网。

①调频网点尺寸：当加网方式为调频加网方式时，设置网点尺寸；

②网点形状阴阳反转：一般只应用于图像文件本身为阴图时，选中此项输出；

③覆盖作业中的网点形状：缺省为选中；

④覆盖作业中的网点频率：缺省为选中；

⑤覆盖作业中的网点角度：缺省为选中；

以上③④⑤三项表示是否用 RIP 中的相关项目的设定值取代待输出文件中的设定值。选中则表示取代。

⑥高于 95% 的黑版叠印：缺省为选中，该选项仅在 RIP 分色时有效。

为了减少印刷中套印难度，避免出现漏白边现象，当黑色达到设置值以上时进行压印，使下面的色版不镂空。

3. 色版打印设置列表

在模板的下部为"色版打印设置列表"，用来设置色版的输出，以及输出该色版时，使用的网角网线数，用户可以选中其中一个色版，通过下拉菜单来设置每个色版的打印参数。

"Default"色版表示除了列表中所列色版外的所有色版。如模板中不添加专色时，则表示除了 CMYK 四色外所有的专色色版，可通过设置此项控制所有专色色版加网参数及输出。

列表右下方有"添加"和"删除"两个按钮,可以通过这两个按钮在列表中添加或者删除一个专色。单击"添加"后,在"色版打印设置列表"中,将自动出现一行数据,选中添加的这行数据,可以在下面的文本框中更改专色名称,以及在下拉菜单中设置该专色的默认打印参数。

"注册当前活件专色"选项是注册当前作业中所包含的专色色版。只针对当前作业进行设置。单击后,流程会将当前作业中包含的专色色版注册到当前模板中。

如果对某专色的加网参数进行单独设置时,在"色版打印设置列表"中添加的专色名必须与文件规范化后文件的"文件信息"对话框中显示的专色名保持一致,否则对专色的设置将不起作用。

"色版打印设置列表"与"高于95%的黑版叠印"只有在分色时才起作用,对于前端已分好色的 PS 文件,输出某个色版与否以及相互之间的叠印关系都已在源文件中描述好,在流程中将无法更改。

4. 专色转换设置项

专色转换是对作业中的专色色版的处理方法进行设置。专色转换处理选项是设置专色色版按照什么方式进行输出,分为"将专色转为 CMYK 输出"和"按活件设置处理专色"。

(1) 将专色转为 CMYK 输出

此参数表示模板中设置输出的所有专色转为四色进行输出。

(2) 按活件设置处理专色

此参数表示专色将根据作业中每个文件的属性——"文件信息"对话框里对专色的设置进行输出。这时,模板只是控制专色是否输出。

第三节 印版输出过程中出现的问题的解决方法

一、图像质量问题及处理方案

输出印版时,若版面上出现质量问题,则引起此问题的原因如下:

计算机:输出印版与计算机的兼容性不匹配、计算机存在病毒;

制版机:焦距偏离、功率不匹配,输送过程中或光鼓上有毛刺,激光不受控,丝杠导轨缺油,机构调整不当;

冲版机:显影温度、速度设置不合理,保养清洁有问题使设备不能正常工作,各辊轴、毛刷压力调整不当,补充液配比不合理;

版材:版材封孔不好、感光胶涂布不均匀,版材的平整度超标,版材的边缘不规则,不适合制版机的版材。

表 4-2 为印版图像质量问题及解决方式。

表 4-2 印版图像质量问题及处理方式

	图像质量问题	原因	解决方案
1	套位不准	1. 左右不准 （1）版左侧定位不准，侧拉规上滚轮压力不够 （2）扫描平台定位不准，丝杠电机联轴节松动 （3）对 C 型机，可能侧拉规螺母和丝杆之间的间隙过大 2. 前后不准或歪斜 （1）进版过程中，输送阻力过大，如送版台上有静电或版材带静电较严重 （2）如有供版器，检查 650mm×550mm 的版材与 1030mm×800mm 的版材的套位精度的差异	1. 调整压力或更换，参照《综合调试流程》 2. 拧紧固定螺钉 3. 确认侧拉规螺母和丝杠之间的间隙小于 0.01 mm （1）用酒精擦导板 （2）供版机的导板与主机的进版高度是否一致。如果小版套位没问题，而大版有问题，侧供版机的摩擦阻力太大，需联系科雷客服部
2	版面有底灰，洗不干净	1. 设备问题 （1）焦距不合适 （2）功率不适合 2. 冲版机的问题 3. 显影温度、时间不合适 4. 冲版机的工作状况不合适 5. 版材问题 （1）版材感度不合适 （2）版材的封孔不合适	1. 确认客户使用的版材合格性检查版材的实际厚度，发调焦图，确定最佳焦距、合适功率 2. 确认冲版机的维护情况 （1）检查实际的显影温度、时间 （2）检查水循环工作压力，水喷淋必须直射，毛刷压力不合适等 3. 检查版的品牌、生产日期 4. 联系版材供应商
3	图像未能发完，扫描平台复位，提示 buffer 空	数据跟不上，数据格式被破坏： （1）计算机在工作时，又有别的进程调入，如数据传递 （2）计算机感染病毒	1. 不要做与作业无关的事情 2. 清除病毒
4	图像局部出现暗线、暗点	1. 版材背面的划痕 2. 光鼓上对应的位置有灰尘	1. 检查使用的版材背面划痕、变形，确认是原有的，还是在输送过程中产生的，找出相对应的位置，清除毛刺 2. 找出光鼓上相对应的位置，清除毛刺
5	版面上有细细的、有规律的白线；图像在实地部分出现白线，或空白部分出现黑线	激光不能有效控制，相对应的激光驱动板或连线有问题	可通过 LaserAdjust 程序把 0 路、8 路设定成 50，发焦距图进行检测，先确定是哪路激光未受控，检查相对应路数的激光品的连线情况，排除接触不良情况 如连线没问题，更换激光驱动板
6	图像上偶尔出现，一段线中，有间隔的白线	主控板有接触不良现象	更换主控板

（续表）

	图像质量问题	原因	解决方案
7	图像上偶尔出现白线	激光驱动的负压配置不对（注：热敏机型）	重新配置负压
8	3%的网点有丢失	1. 激光功率偏高 2. 冲版机的冲版温度及时间有变化	1. 确定合理的曝光功率 2. 检查冲版机的温度、时间。需要注意的是，补充液是否按规定进行补充，补充液的合理配比，建议补充液的配比应与显影液的配比相同
9	98%的网点中，白点不明显，有条纹印	激光功率偏低	在确保冲干净的情况下，调整激光功率
10	整个版面上有花纹，或有刷痕	冲版机的各胶辊压力未调整到位，毛刷辊的压力未调整到位	调整各胶辊的压力、毛刷辊压力
11	图像上有淡淡的条纹或带脏，冲不干净	显影不干净，显影液失效或冲版机条件设置不当	可适当提高药水温度或降低冲版的速度，或更换药水
12	发版时，偶尔有图像头部错位	锁相不稳： （1）码盘信号被干扰 （2）伺服电机运动不稳	检查有效接地情况
13	发版时，偶尔有局部乱码，并且有时会超出图文部分	1. 文件被损坏 2. 码盘信号被干扰	1. 文件预览时，显示情况是否正常 2. 由于计算机主板与内存的不兼容性，也会造成文件被破坏。可先更换一台无病毒的计算机 3. 测试，确定文件在制作过程中未被破坏 4. 检查有效接地情况
14	图像中，满版的小白点或小黑点	版材问题	可换一种品牌的版测试
15	图像有痕迹，非激光扫描造成	1. 版材问题 2. 冲版机胶辊涨	1. 因版材本身药膜较嫩，检查在输入或输出时，是否有摩擦力造成版面的痕迹 2. 检查冲版机各辊系，是否有对版面造成损伤
16	出版有黑线，分不规则黑线及间隔4mm、5mm黑线	1. 丝杆、导轨缺油 2. 扫描平台对光鼓的直线度有偏差 3. 丝杆局部的运动阻力有变化	1. 按要求重新清洗丝杠、前后导轨，并加润滑脂 2. 用千分表检查，光鼓和扫描平台的直线度小于0.006mm；如有超差，则需调整平行度 3. 用千分表检查丝杠的运动精度，其应小于0.004mm，如果超出需重新研磨丝杆

二、其他异常情况及处理方式

表 4-3 为系统工作异常及处理方法。

表 4-3　系统工作异常及处理方法

序号	异常情况	原因分析	处理方法
1	进版不顺	1. 进版真空吸附力不适合（D/E 型机） 2. 进版滚轮压力不适合（C 型机） 3. 供版机与主机的左右高低位置不符 4. 版材或导板上有静电	1. 调整真空吸附力，见综合调试流程 2. 执行 LaBoo 程序，运行测试程序 151 命令，调整两组滚轮压力一致，调整时以 0.15mm 厚的版材为基准，版送至版头前规时，遇到阻力，版材中间不拱起 3. 供版机的送版导板与主机的进版导轨高低一致，左右居中，当版材送到主机时，版材左边缘与侧拉规指针有 3mm 以上的距离 4. 消除静电，可用酒精擦洗导板，版材上有静电需消除
2	卸版不顺	1. 卸版版头打开的角度不够 2. 使用的薄版版基太软	1. 调整打开角度 2. 与公司方客服部联系
3	退版时卡版	1. 可能装版时，版材已经歪斜，退版时，碰到侧拉规 2. 对 C 型机，退版导板安装有问题	1. 检查输送动力及输送是否歪斜 2. 退版导板必须与进版导板平行
4	版尾调整失败	1. 光鼓版尾位置定位不准确 2. 版尾打开时角度不够，未能有效把锁齿脱开 3. 版尾左、右扭簧损坏 4. 版尾摇臂拉簧损坏	1. 检查 3 号参数是否准确 2. 检查版尾开闭执行的步进电机、驱动器，传感器是否工作正常，调整 5 号参数 3. 更换版尾左、右扭簧 4. 更换版尾摇臂拉簧
5	版尾斜度超标	1. 版材不规则 2. 0 号参数设置不当	1. 检查版材的角尺，确认版材符合标准；在中间部 200 mm 内，允许误差 ±4 个数字，约 ±0.4 mm 2. 重新检查 0 号参数
6	版材长度与模板不符	1. 版材的实际长度不标准 2. 8 号参数设置不当	1. 检查版材的实际长度，允许误差 ±10 个数字，约 ±1 mm 2. 重新设置 8 号参数
7	报"光鼓上无版"或"光鼓上有版"请执行命令卸版	光鼓上有无版检测探头调整不当	执行 105 命令，使"无版检测探头"复位
8	曝光过程光鼓速度异常	1. 计算机病毒攻击程序 2. 光鼓皮带偏松，或光鼓皮带磨损严重 3. 电源质量不稳定	1. 清除病毒 2. 调整皮带的松紧度 3. 确保电源质量在 220AC（1±15%）内
9	激光箱温度超出工作范围（在环境温度满足的前提下）	1. 测温传感器（18B20）损坏 2. 恒温驱动板可能损坏或线头接触不良	1. 确认温度传感器工作状态，可以和机架温度传感器互换查找原因 2. 用万用表测量恒温驱动板的 4P 插头，正常状态下，输入电压 15V，输出电压约 9V，如果一直偏高或偏低，则应重新定标

（续表）

序号	异常情况	原因分析	处理方法
10	锁光失败	1. 出光未对准能量检测传感器探头中心 2. 镜头、密排、能量检测传感器表面有灰尘 3. 连线的线头接触不良 4. 激光驱动板有问题 5. 光纤与激光器耦合不良 6. 光纤损坏 7. 激光器损坏	1. 检查对准情况 2. 对各表面的灰尘进行清洁 3. 检查激光器的电压是否正确 4. 检查驱动板是否有问题 5. 检查光纤的耦合处是否漏光 6. 检查密排的光纤是否漏光 7. 更换激光器 可通过互换的方法，确定是哪个零件有问题
11	平衡块调整失败	1. 版材是否为标准规格 2. 电机、驱动器、传感器工作是否正常 3. 传感器安装是否正确 4. 参数是否正确，定位、滑动齿打开是否正确 5. 机构的各零部件固定是否准确	1. 确认版材是标准的，制版后，不用裁切 2. 确定电机、驱动器、传感器工作正常 3. 确定复位正确 4. 光鼓定位准确，锁齿能有效打开 5. 确认各零部件运转正常

第五章　印版的显影

学习目标

1. 判断显影条件并设置显影参数；
2. 能用检测仪器检测显影液的酸碱度（pH）、温度和电导率；
3. 能保养显影机等制版设备。

第一节　CTP 印版的显影

一、印版显影基本操作

印版显影的基本操作如下：通电开机，首先利用操作界面设定必要的工艺参数，如显影温度、电机速度、版材尺寸规格等。在工艺参数设定完成以后，选择相应的程序让机器进入工作状态。机器进入加温状态，与此同时，仔细观察循环泵的工作正常与否，具体可看液体的流动状态或用温度计测量显影温度是否上升。

当补充液使用完又未及时更换新的补充液桶，而导致显影槽药水液位偏低，换新补充液后，可在"手动操作"内用"Dev-Rep"吸一些药水进去，保证显影槽溢满。冲版机水槽水位感应器存在感应失灵现象，在已加满水后若仍报"Water tank is not full"（水槽不满）错误，用物体接触感应器几下即可。机器用的冲版程序如不小心按到其他程序，会提示是否更改，按"NO"，再在出现的新界面上按下"ESC"即返回主页。

需要注意以下几项。

（1）急停操作：按下"急停"后机器即停止转动，再拔回，然后按绿色按钮，再按报错界面边角处的"主页"键即返回主页界面。

（2）卡版处理：先急停，然后松动胶辊等构件后取出。

（3）如果版在前方平台停滞了一段时间，再推入后会出现中途停转现象，此时机器会自动报错，会发出急促警示响声，应快速按下报错界面上的"主页"键，再把手指放到前方感应器，让机器重新运转起来。这种情况一定要立即处理，因版不能在显影槽中停时间过长。

（4）螺栓注意上紧。

二、注意事项

（1）确认设备使用状况，如果是没使用过的新设备填写表5-1之后可直接进行步骤9，否则放药准备清洗。

（2）确认客户CTP设备的型号规格以及其他相关信息，并填写表5-1。

（3）将显影槽（水洗槽如果有）以及胶槽排空并关紧阀门，把滤芯水放净并摘除滤芯，摘除冲版机进版，显影，水洗以及上胶的胶辊和毛刷待清洗，注意做好记录记好顺序，用干净水将三个槽都灌满冲洗至少两遍，中间需用抹布或者其他清洁用品擦洗揉搓，注意不要将水冲洗到电路部分造成短路。

（4）将显影槽灌满水，按照比例配入显影机清洗剂搅匀循环2小时以上（温度设定30℃）。

（5）用显影机清洗剂原液直接擦洗拆下的胶辊，将胶辊清洗至露出原来橡胶的原色，整根胶辊两头比较难清洗但是也一定要洗干净。

（6）显影机毛刷如果有聚合粘黏物也必须清洗干净，可以采用清洗剂浸泡法或者在保护措施到位的情况下可以用毛刷蘸取显影机清洗剂原液刷洗毛刷，但清洗后需要用显影液中和酸性。

（7）显影槽浸泡时间完成后排空清洗剂，用清水灌满冲洗两遍及以上，中间需用抹布或者其他清洁用品擦洗揉搓，遇到顽固部分要费力擦洗，注意不要将水冲洗到电路部分造成短路。

（8）将显影机内清洗设备的水全部排空，安装好胶辊和毛刷，灌入显影液。

此时应注意显影机内管道残存的水对显影液造成的影响，应在上药水时先将循环回路两端关闭，待少量显影液灌入显影槽时开启循环回路的一端，利用槽内的显影液将管道内的水排到过滤器内排掉，再关闭已经排完水的一端开启另一端进行相同的操作，水排除干净以后将过滤器摘下安装过滤芯。

显影槽上液完成后可以升温调整显影条件，准备测试并将测试结果填写至表5-1。

表 5-1　印版测试记录

公司全称				
制版情况				
制版机	型号：		数量：	
	上版方式：		曝光方式：	
使用条件	能量值：		转速：	
使用版材	型号：		批号：	
	规格：		月均用量：	
显影情况				
显影机	型号：		容量：	
显影液	型号：		批号：	
	冲版量：		使用时间：	
	补液用量：			
显影条件	温度：		速度：	
	动补：		静补：	
保护胶	品牌：		配比量：	
烤版胶	品牌：		配比量：	
使用情况				

一、反映问题
二、情况了解
三、问题处理

第二节　CTP 印版的显影问题解决

一、电导率传感器及控制板的安装调试说明

1. 单台冲版机配件

单台冲版机配件需求如表 5-2 所示。

表 5-2　冲版机配件

序号	配件名称	数量
1	电导率仪	1
2	电导率仪支架 1	1
3	电导率仪安装螺母	2
4	电导率控制板	1
5	电导率控制板套线	1
6	继电器	2
7	护线圈	1
8	（12.9S）内六角螺钉 M3*16	2
9	（304）大扁头十字螺钉 M4*12	1
10	防松螺母 M4	1

2. 安装说明

把冲版机上原安装液位仪和温度传感器支架更换为电导率仪支架，具体安装步骤及安装要求如图 5-1～图 5-15 所示。

（1）首先必须关掉冲版机电源，而后拆卸电气箱盖及后盖（见图 5-1）；

图 5-1　拆卸电气箱盖及后盖

（2）打开冲版机电源，了解冲版机补液泵未进入补液状态下的正常指示灯情况，然后进入"手动操作"功能下进行强制补充，进而根据指示灯情况了解控制补液泵的继电器，以确定补液泵的正常情况和电源线路（见图 5-2）；

（3）再次关掉冲版机电源，从"电导率控制板套线"中找出用于连接"冲版机卡板上 220V 火线"和"电导率控制板 24V 电源线"的线，并从冲版机线槽中穿出，至电导率传感器和继电器的安装位置，与电导率控制板控制线连接（见图 5-3、图 5-4 和图 5-15）；

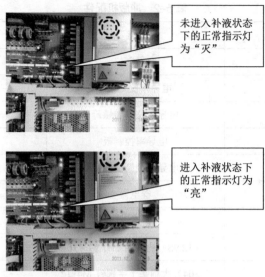

图 5-2 补液状态下的指示灯

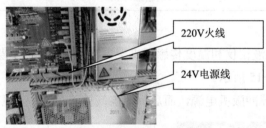

图 5-3 火线与电源位置

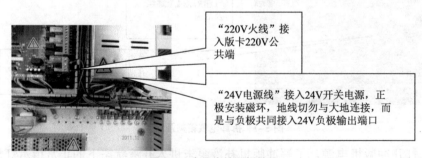

图 5-4 电导率控制板接线

（4）拆卸冲版机右罩（见图 5-5）；

图 5-5 拆卸冲版机右罩

（5）冲版机自带补液开关上三芯接插件上的火线剪短，输入输出两端分别接出从"电导率控制板套线"中找出的两根"220V火线分线"（见图5-6和图5-15）；

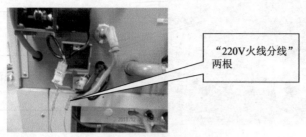

图5-6　电导率控制板套线

（6）继电器的安装：采用工具"博世带扭矩电钻""钻头Φ2.5""丝锥M3"打出两个用于固定继电器的M3螺纹孔，孔深约5mm，间距约20mm，切勿打穿冲版机药槽侧板，再用配件"（12.9S）内六角螺钉M3×16"和工具"2.5内六角扳手"固定继电器，注意螺丝松紧度适中，切勿死拧，因为药槽侧板为PVC板，死拧易滑牙（见图5-7）；

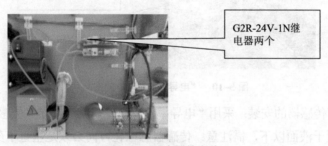

图5-7　继电器的安装

（7）"220V火线分线"分别接入内外部控制切换用继电器"公共端"和"常闭端"（见图5-8和图5-15）；

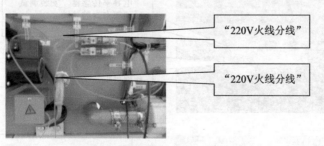

图5-8　继电器"公共端"和"常闭端"

（8）工控继电器输出线的安装：从"电导率控制板套线"中找出的一根"K1内部/外部控制切换"输出电源线1和"K2补液开关"输出电源线2分别与两个继电器连接，另一端电导率控制板控制线连接（见图5-9和图5-15）；

（9）"电导率支架1"的安装：采用工具十字螺批、呆扳手7mm×9mm、活扳手6寸将原安装温度传感器和液位仪的支架拆卸下来，更换为"电导率支架1"（CTP-GZ-157A），注意液位仪的拆卸需将连接线从连接端子处拔出后从药槽侧板孔拔出才能卸下（见图5-10）；

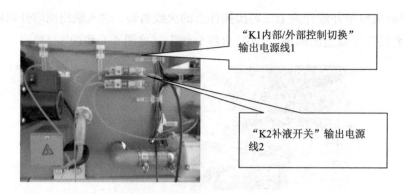

图 5-9　工控继电器输出线的安装

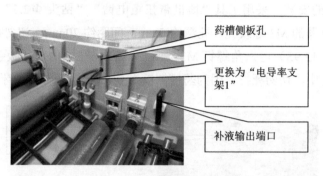

图 5-10　"电导率支架 1"的安装

（10）电导率传感器的安装：采用"电导率仪安装螺母"将电导率传感器安装于支架上，电导率检测球须置于液面以下，需注意：传感器感应孔方向必须是垂直于药槽侧板方向，即与胶辊方向平行，否则易造成由于药液频繁流动而使传感器监测不稳定的现象（见图 5-11）；

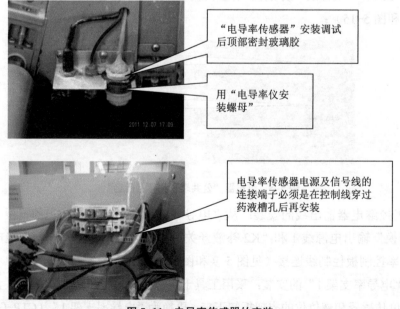

图 5-11　电导率传感器的安装

（11）电导率控制板的安装：采用工具"博世带扭矩电钻""钻头 Φ3.2""丝锥 M4"打出"电导率控制板"悬挂螺丝孔 M4，采用零件"（304）大扁头十字螺钉 M4 12"安装，内侧紧定"防松螺母 M4"；采用工具"博世带扭矩电钻""钻头 Φ3.2""钻头 Φ13.5"打出"电导率控制板"连接线穿孔，同时采用"护线圈"安装孔内，以保护电导率控制线（见图 5-12）；

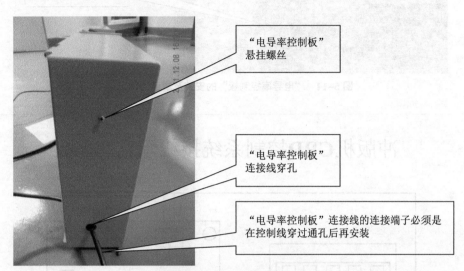

图 5-12　电导率控制板的安装

（12）"电导率控制板"连接线与各电源线、控制线、信号线的连接：注意各线路的固定（见图 5-13 和图 5-15）；

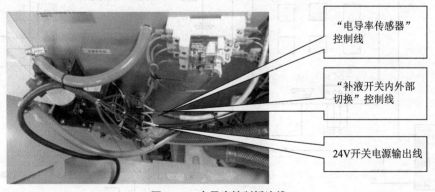

图 5-13　电导率控制板连线

（13）"电导率控制板"的安装及罩壳的固定：恢复冲版机的电气箱盖和外罩壳（见图 5-14）；

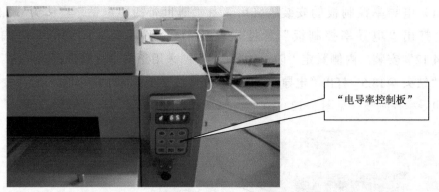

图 5-14 "电导率控制板"的安装及罩壳的固定

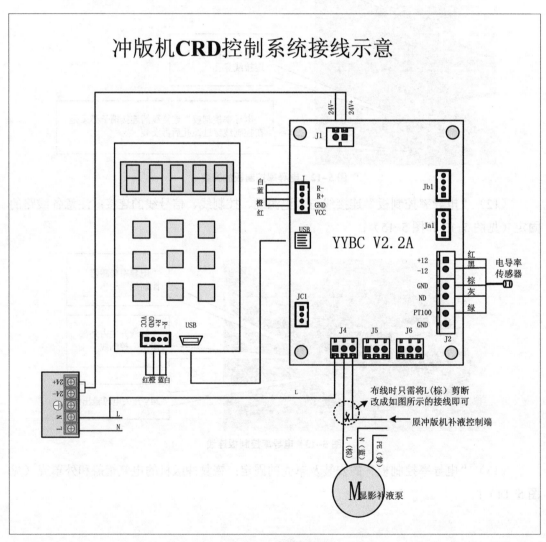

图 5-15 电路控制

二、印版显影参数设置

电导率控制显示面板的显示界面中,第一位为参数项,后四位为参数值。

1. 菜单键

在显示模式下,短按该键将进入设置界面。在设置模式下,短按该键将切换到下一个参数设置界面。在任意模式下,长按该键 5 秒将切换到氧化补偿设置界面。

2. UP 键

在设置模式下,显示屏幕不闪烁时,短按该键将切换到上一个参数设置界面。屏幕位置闪烁时,短按该键将增加该位值。在任意模式下,长按该键 5 秒将清零低于下限值情况下的补液次数。

3. TURN 键

在设置模式下,短按该键将切换闪烁位,以修改设定参数。

4. ENTER 键

在设置模式下,短按该键将保存当前参数(恢复出厂值等特殊功能除外)。在任意模式下,长按该键 5 秒将初始化计时芯片。

5. DOWN 键

在设置模式下,该位不闪烁时,短按该键将切换到下一个参数设置界面。该位闪烁时,短按该键将减小该位值。在任意模式下,长按该键 5 秒将切换报警功能开启与关闭。

6. ESC 键

在设置模式下,短按该键将退出设置模式并显示等待界面,5 秒后显示电导率界面。

7. F1 键

在补液状态下,短按该键将关闭当前补液状态。在任意模式下,长按该键 3 秒将进入强制补液状态,直至电导率到达设定值关闭。

8. F2 键

在任意模式下,长按该键 3 秒将切换补液控制方式(CRD/冲板机)。

第六章 印版质量的检测

 **学习目标**

1. 能及时发现印版的划伤、折痕、脏痕等缺陷；
2. 能对照签样检查版面尺寸，有无丢字、乱码、缺图和变形等问题；
3. 理解印版测量仪器的原理；
4. 能用测量仪器测量印版的网点及角度；
5. 能检测加网文字、线条的清晰度和完整度；
6. 能借助测量仪器和测控条，检查网点形状完整性和网点增大值，并提出制版工艺的改进建议；
7. 能对印版质量进行综合检查，对产生的问题提出解决方案。

第一节 CTP 常见质量问题

一、印版在 CTP 中的检测

①图像要呈居中状态；②咬口与印刷的要求一致；③版尾压边宽度要达到 5～6 mm，可确保印版在光鼓高速运作中不飞版；④网点还原率在 1% 内；⑤印版要求无底灰；⑥四色套印要求在 0.01mm 以内。

二、常见的印版缺陷

源文件和生成的 TIFF 文件之间的内容不同，不同的表现有图文丢失、图文变形等。人工目测检查较为困难，一般通过一些自检软件进行自动对比，如 TellRight 等预检软件，在美工、拼版、Rip 等各个环节进行自行检测。

外物造成印版划伤、折痕，主要在包装、运输及冲版过程中可能会产生。冲版划伤有可能是胶辊、毛刷老化或结晶较多产生的刮痕。外物造成的印版伤痕大多无规则。

激光出错或振动也会对印版造成伤痕，该伤痕表现是版头到版尾的垂直的线，是有规律的。

能量不足、显影不足或显影液失效也会对印版造成伤痕，该伤痕主要体现在满版出现斑斑点点，呈现感光胶的颜色。

第二节　印版测量仪器

一、功能说明

Techkon Spectro Plate 测版仪出色的图像捕捉质量和独创性的图形算法，能够精确读取任何加网尺寸和加网技术：调频网、调幅网或者混合加网。白色照明光谱和动态色彩评估使得设备能够读取所有类型的印版和表面涂层。测版仪（见图 6-1）的优势不仅仅在印版识别上。这个多功能设备在进行胶片的网点测量和 CMYK 四色印刷纸测量时同样出色。

图 6-1　测版仪（Spectro Plate）

二、测量网点类型设定和测量方法

根据加网类型：调幅网 AM、调频网 FM，选择对应的设置，将仪器放在印版测控条需要测量的位置上，按绿色测量按钮，在仪器上即可看到对应的数据：网点面积，如果是调幅网还可以看到加网线数、加网角度。调频网 FM 及调幅网 AM 测量功能分别如图 6-2、图 6-3 所示。

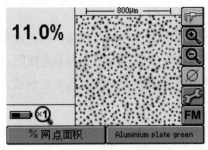 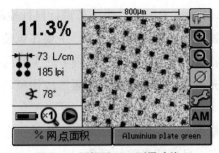

图 6-2　调频网 FM 测量功能　　　　　图 6-3　调幅网 AM 测量功能

Techkon Spectro Plate（见图 6-4）可以在 PC 上控制，显示仪器的操作界面，并能保存仪器截图。Techkon Spectro Plate 可以连接到计算机，操作设备测量时，会在计算机上显示仪器的屏幕信息。

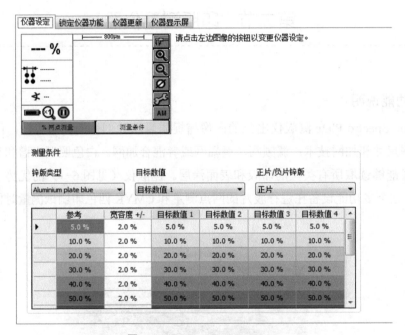

图 6-4　Techkon Spectro Plate

第三节　印版输出检测

一、确定版材的最佳焦距

1. 焦距的定义

光学系统中衡量光的聚集或发散的度量方式，指平行光从透镜的光心到光聚集之焦点的距离。

简单地说,焦距是从镜头的镜片中间点到光线能清晰聚焦的那一点之间的距离。

2. 变焦的定义

曝光时对于焦点和焦距的相应调整。

对焦的定义:调整焦点,使被扫描图像位于焦距内,成像清晰。

(1)正常的焦距

在原有正常工作的功率和冲版条件下输出一张带有科雷的焦距测试图的版子,把版子版尾边靠近人的身体放在桌子上或者看版平台上,肉眼观察焦距测试块中的焦距线都应全部均匀排列,没有条纹,如图 6-5 所示。

图 6-5　焦距线

(2)物距对焦距的影响

在焦距测试时,把第 0 路激光关闭,关闭的激光会在焦距图中呈现明显的黑线,每根线条代表一路激光的焦距线。如图 6-6 所示,物距偏小时,通过放大镜观察黑线的延伸线比周边其他线的颜色要深些,俗称"漏黑"。物距偏大时,通过放大镜观察黑线的延伸线比周边其他线的颜色要浅些,俗称"漏白"(见图 6-7)。

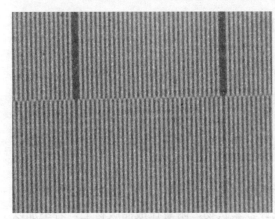

图 6-6　物距偏小的焦距线

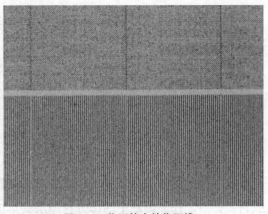

图 6-7　物距偏大的焦距线

（3）最佳焦距的判定方法

方法一：用放大镜在焦距测试图上从右往左观察，第 M 格出现 1×1（像素 × 像素）网点，过了 N 格后 1×1（像素 × 像素）网点消失，焦距的最佳位置应是 M+N/2 偏左一格。

方法二：选择能还原出网点的最小的网点区域，如 1% 的网点区域已经能还原出网点，选择放大镜从右往左观察 1% 的网点区域，第 M 格能清晰看到网点的还原，在第 M 格之前网点丢失严重；到第 N 格时网点呈现还比较清晰，但第 N 格之后网点明显出现丢失现象，此时焦距的最佳位置应是 M+N/2 偏左一格。

二、确定版材的最佳曝光条件

不同品牌版材所要求的最佳曝光和冲版条件是不尽相同的。以黑木 UV 版材为例，适用黑木显影液，假如 UV 版材厂家提供标准药液配比 1∶4，显影液电导率 60ms/cm，毛刷转速 100r，冲洗温度 25℃，显影时间 20s。

输出一张变功率测试图，通过滴丙酮的检测方法，检测丙酮溶液在无图文的版材上扩散是否留有底灰，扩散有蓝色圆圈状的现象视为有底灰，扩散呈白色状态的现象视为无底灰，在无底灰区域用网点检测仪进行检测，实测 50% 的网点区域为 50% 或者小两三个点的网点还原，同时又保证没有底灰，就可以选择略小于 50% 的网点还原的功率值为最佳功率。若选择的最佳功率较大，接近设备的功率上限值，可以通过增加显影液时间和温度的方式，调低最佳功率。

第四节　CTP 印版成像质量控制

一、影响 CTP 印版成像质量的主要因素

1. 版材质量

不同类型的版材的成像质量是不同的。对于一个企业来说，最好选用固定厂商、固定型号的 CTP 版材，这样可提高制版质量的稳定性。

2. 设备性能

不同设备的曝光性能不同，单位面积光源照度和均匀度会对网点的均匀性产生影响。

3. 显影条件

显影液的化学成分、温度、浓度等都是影响制版质量的关键因素。同时，一定数量的版材显影后，在显影液中的部分树脂层会形成许多絮状物，附着在成品版材上，若不加以处理会造成印刷时带脏。

4. 工艺控制

工艺控制主要指各种工艺参数的设置，如曝光时间、显影时间等。另外，同一套版最好一次性处理完，这样能保证套印精度。

5. 环境条件

环境条件主要指制版车间的温度、湿度、光照条件等。应设定在版材所要求的范围内。

6. 后处理工艺

印版出现个别点状感光胶残留时，可用修版笔进行修复。

二、CTP 印版的质量控制方案

CTP 制版进行质量控制的基本前提是调节好制版设备，使整个计算机直接制版系统处于最佳状态。

曝光和显影是计算机直接制版过程中最重要的程序，因此调节制版设备主要是针对曝光参数和显影工艺的控制。

（1）制版机曝光参数的控制

要用好 CTP，首先就要控制好制版机的曝光参数，使它的光学系统和机械系统处于良好的状态。在用户拿到一款和制版机曝光机制适应，波长范围匹配的版材后一定要对版材进行感光性能测试。测试项目包括激光焦距与变焦测试（FOCUS/ZOOM TEST）、激光发光功率和滚筒转速测试（LIGHT/ROTATE）。其中激光焦距与变焦，功率与转速可以做组合测试。

制版机通常带有自己内部的曝光参数测控条，通过测控条上的色块或者图案可以很方便地检测印版曝光量是否合适、激光头聚焦是否正确等硬件设备的状态。

（2）冲版机显影工艺的控制

印版正常曝光之后，还需在冲版机里面进行正常的显影才能得到模拟图像。因此必须对冲版机的状态进行测试和监控。

随着使用时间的延长，硬件设备都会出现衰老，设置值和实际值之间会存在一定差异。必须对冲版机的状态进行测试。测试时，需要利用显影液专用温度计对冲版机的"实际温度"进行取点测量。利用大量筒或者量杯对冲版机"实际动态补充量"进行计时计量监控。若是设置值与实际值差异太大，则需改善循环系统或更换传感器件。

当冲版机硬件状态监控好后，则需进行显影液匹配测试。用户可以利用各大公司的标准数字印版测控条进行测试。另外，也可自制印版控制条进行测试。利用数字印版控制条，可以分析印版上的网点的变化情况，从而判断印版是否正常冲洗（印版曝光正常为前提），显影温度和显影速度的参数设置是否正确。一般情况下正常 CTP 印版上 2%～98% 的网点都应该齐全，50% 的网点增大不超过 3%，95% 的网点不出现糊版、并级现象。

三、利用数字制版测控条进行数字监控

CTP 系统是数字化工作流程中的一部分，所以数字化控制方法对质量保证是必不可少的。数字制版控制条可以对 CTP 印版的成像质量进行合理有效的控制。

用于 CTP 印版控制的数字测控条主要有 GATF 数字制版控制条、Ugra/Fogra 数字制版控制条、柯达数字印版控制条、海德堡数字印版控制条等。其中使用最广泛、最主要的是 Ugra/Fogra 数字制版控制条和 GATF 数字制版控制条。

（1）Ugra/Fogra 数字制版控制条

该控制条中包含六个功能块和控制区，如图 6-8 所示。

图 6-8　Ugra/Fogra 数字制版控制条

①信息区：包括输出设备名称、PS 语言版本、网屏线数、网点形状等。

②分辨率块：包含两个半圆区域。线条自一点发出，呈射线形排列，射线的浓密度与输出设备理论上的分辨率一致。在线条中心形成一个或多或少，敞开或封闭的四分之一圆，这两个四分之一的圆越小和越圆，聚焦和成像的质量越好。左边为阳线，右边为阴线。

③线性块：由水平垂直的微线组成，用来控制印版的分辨率。

④棋盘区：由 1×1、2×2、3×3 和 4×4（像素 × 像素）构成的棋盘方格单元。控制印版的分辨率，显示曝光和显影技术的差异。

⑤视觉参考梯尺（VRS）：控制印版的图像转移。

⑥网目调梯尺：主要用于通过测量确定印版阶调转移特性。同时所提供的 1%、2%、3% 和 97%、98%、99% 的色块也可用于对高调和暗调区最终所能复制出的阶调进行视觉判断。

其中，视觉参考梯尺（VRS）是 Ugra/Fogra 数字制版控制条的一个特殊之处。它是进行图像转移控制的基本要素，控制印版的稳定性，使数字印版的生产程序标准化。在 VRS 中包含有成对的粗网线参考块，在其周围则是精细加网区域。控制条中共有 11 个 VRS，并且在从 35%～85% 的网点区域里按 5% 的增量递增。在理想状态和线性复制的情况下，VRS4 中的两个区域在视觉上应该具有相同的阶调值。但是实际上，两个区域具有相同阶调的 VRS 要比 VRS4 高或低，这取决于印版类型和所选的校准条件。VRS 是一个非常理想的过程控制块，利用它无须进行测量，直接从视觉上就可判断出与所选条件的差别，进行视觉检查。

（2）GATF 数字制版控制条

GATF 数字制版控制条，如图 6-9 所示。

①信息区：包括输出设备名称、PS 语言版本、网屏线数、网点形状等。

②阳图阴图的水平垂直细线：测试系统的分辨率，控制曝光强度。

③棋盘区：由1×1、2×2、3×3和4×4（像素×像素）构成的棋盘方格单元。

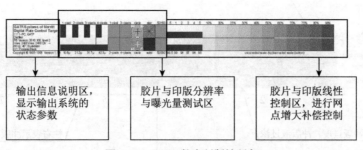

图6-9　GATF数字制版控制条

④微米弧线区：阳图和阴图型微米弧线。使用设置的最小尺寸以弧线段对系统检测，微米弧线图案是对系统最严峻的挑战。如果一个系统同时保持对阳图和阴图弧线的良好细节，就表明该系统具有良好的曝光条件。

⑤星标对象：测试系统的曝光强度，分辨率和阶调转移特性。

剩余的部分是两套匹配阶调梯尺。两个阶调梯尺的不同之处在于，上面一个绕过了应用于其他文件的 RIP 的补偿程序，而下面一个则没有绕过补偿设置。对两个梯尺的比较清楚的表明补偿程序所造成的影响。使用阶调梯尺，首先要用放大镜观察图像系统的高光和暗调的限定，然后使用密度计从10%到90%的测量阶调梯尺，从而构建网点增大值曲线。

（3）科雷的数字制版测控条

科雷标准数字制版测控条一般包含五个功能区域，如图6-10所示。

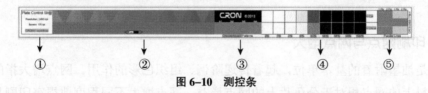

图6-10　测控条

①信息区：包括输出设备名称、PS语言版本、分辨率、网线数等。

②显影效果区：见图6-11，该区域分为七块方块区域，通过方块内倒三角区域与方块内其他区域颜色的对比，颜色最接近的方块处于3—5区间内，表示显影效果良好。颜色最接近的方块处于1—2区间内，表示冲版过度，需要调整显影条件。颜色最接近的方块处于6—7区间内，表示显影不足，需调整显影条件或者更换药水。

③像素区：由1×1、2×2、3×3和4×4（像素×像素）构成的圆形区域，检测印版分辨率和对应精度激光对焦的效果。

④网点还原区：由0%、1%、2%、5%和50%、97%、98%、99%的网点区域组成，主要检测CTP设备小网点和大网点的还原效果，也可用于对高调和暗调区最终所能复制出的阶调进行视觉判断。该区域又分上下两块区域，下面一个绕过了应用于其他文件的RIP的补偿程序，而上面一个则没有绕过补偿设置。对两个梯尺的比较清楚地表明了补偿程序所造成的影响。

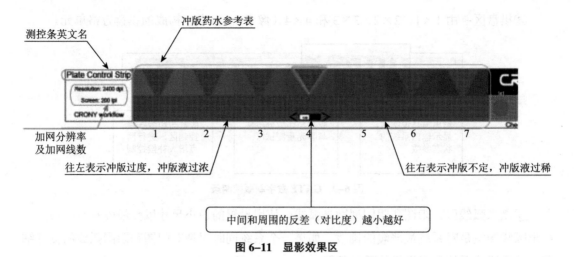

图 6-11 显影效果区

⑤线条区域：由 1×1、2×2（像素 × 像素）的水平线和斜线组成，可以检测印版分辨率，也可以用来判断图像起始曝光点位置是不是最佳位置。

第五节 网点增大的原理及控制

一、印刷网点与网点增大

网点是油墨附着的基本单位，起着传递阶调、组织色彩的作用。网点增大指的是印刷在承印材料上的网点相对于分色片上的网点增益。网点增大不同程度地损害印刷品，破坏画面平衡。但是由于技术和光线吸收的原因，没有网点增大的印刷是不存在的，如图 6-12 所示。

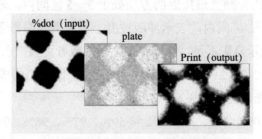

图 6-12 网点在印刷工艺中传递的状态

印刷生产控制的目标之一就是为所有印刷机按纸张分组而规定相应的网点增大标准，并在制作胶片时考虑这个网点增大标准值，从而通过工艺补偿对网点增大进行控制，实现印刷图像色彩与阶调复制的理想结果。为了准确地获得网点增大标准，需要对印刷网点的增大进行测量。

网点增大的测量通常是在具体的印刷材料、设备器材和理想印刷压力条件下，用晒有网点梯尺与包含有实地、50%、75% 的网点内容测试条（如布鲁那尔测试条）和任一图像画面的印版印出数张样张。印刷时要保证每张样张网点整洁、实在、无重影变形。然后，用反射网点密度计分别测出各样张的四色实地密度，50% 与 75% 处的印刷网点面积率。最后将印刷网点与 50% 或 75% 求差值，即可测量网点增大值。

二、网点增大计算（TVI）

1. 基于密度值的网点增大

在实际印刷过程中，无论用哪种方法来计算颜色，都应该先对网点增大进行修正。网点增大是对印刷颜色影响最大的因素。通常，在计算印刷网点面积和网点增大量时使用密度计算法，也就是要使用玛瑞-戴维斯（Murray-Davies）公式：

$$a = \frac{1-10^{-D_t}}{1-10^{-D_o}} \qquad (6-1)$$

或使用尤拉-尼尔森（Yule-Nielson）公式：

$$a = \frac{1-10^{-D_t/n}}{1-10^{-D_o/n}} \qquad (6-2)$$

其中，a 是色调密度为 D_t 时的单色原色油墨网点面积率，D_o 为印刷实地密度。因此，用密度值控制印刷的条件非常方便。计算网点增大，则再使用定义的值与 a 计算的结果相减就可以得到。

2. 基于三刺激值的网点增大

ISO/TC10128 标准规定，网点增大值（Tone Value Increase，TVI）可以通过测量一系列色阶的三刺激值，结合实地四色三刺激值计算得到。

$$\text{黑色与品红色网点增大值} = 100\left(\frac{Y_P - Y_t}{Y_P - Y_S}\right) - \text{TV}_{\text{Input}} \qquad (6-3)$$

$$\text{黄色网点增大值} = 100\left(\frac{Z_P - Z_t}{Z_P - Z_S}\right) - \text{TV}_{\text{Input}} \qquad (6-4)$$

$$\text{青色网点增大值（TVI）} = 100\left[\frac{(X_P - 0.55Z_P)-(X_t - 0.55Z_t)}{(X_p - 0.55Z_p)-(X_S - 0.55Z_S)}\right] - \text{TV}_{\text{Input}} \qquad (6-5)$$

其中，X_P、Y_P、Z_P 是纸张的三刺激值；X_S、Y_S、Z_S 是实地青色、品红色与黄色、黑色的三刺激值；X_t、Y_t、Z_t 是不同阶调的四色色阶三刺激值。

三、制版过程中避免网点增大的方法

网点是表现印刷品层次、阶调和色彩的基本单位，印刷网点的变化往往会导致色彩还原失真以及阶调层次减少等质量问题。印刷流程中常见的网点转移问题包括网点的增大、滑移、重影等，其中最常见的问题就是网点增大。导致网点增大的主要因素有两大方面：照排机或 CTP 的非线性特点以及曝光系统的阶调传递特性等因素；印刷机、纸张和油墨的类型以及印刷压力等印刷工艺条件。

1. 输出阶段的网点增大补偿

理想条件下，照排机或 CTP 制版机接收到 PS 或 ONE-BIT-TIFF 文件后，输出的激光量应该与文件上不同图文部分的网点面积成正比。电子图像的每 1 个像素被输出为 1 个网点，像素的灰度值与网点大小是一一对应关系。但是，由于机器制造的工艺精度等原因，大部分设备实际输出效果并非如此，往往会呈现出一定程度的非线性，再加上其他光学以及显影方面等因素的影响，就会造成软片或 CTP 印版上输出的网点图像的阶调偏离原像素值。

为了获得需要的网点大小，必须采取措施修正偏差，使像素值与最终的输出保持线性关系。这种偏差的修正可以利用 PS 页面描述语言中的变换函数实现。这一补偿需要在 RIP 之前进行，通过修正 PS 分色文件达到对制版阶段产生的网点扩大做出补偿的目的，这一过程一般有以下两种实现手段。

（1）通过制版流程软件线性化进行补偿

RIP 制造商在其产品中提供了自定义变换函数的功能，即可以在其 RIP 中调用 PostScript 语言的变换函数操作符，这种网点增大补偿的方法通常称为线性化。在新照排机正式投入生产之前必须进行线性化调整；在更换不同的显影液或胶片之后也应当做当前状态下的线性化。

对 CTP 制版机而言，在改变印版或冲版条件以及改变制版机的 FOCUS、ZOOM 或激光功率值时，要做到精确的网点控制，也要相应地重新做该条件下的线性化。线性化补偿的具体做法见前面相关章节。

（2）利用 Adobe Photoshop 软件的传递函数

因照排机标定不当而导致的网点增大，除了输出设备线性化的方法之外，利用 Adobe Photoshop 软件内置的传递函数功能，建立自定义的传递函数曲线，也可以达到补偿这个目的。

在 Adobe Photoshop 软件中利用网点传递函数进行网点增大补偿的具体操作如下：

①由照排机输出检测片，利用透射密度计测量胶片上不同网点面积率的密度值；

②执行文件菜单中的"页面设置/传递函数"命令；

③计算需要的调整值，将计算得到的值输入传递函数对话框相应的方框内。所谓的计算是指假如指定输出 50% 的网点，照排机输出的网点为 52%，这样输出的网点增大值为 2%。为了补偿这一增大值，应在该对话框 50% 文本框内输入 48%。这样，在调用包含这一转

移函数的分色文件输出时,就会得到所需要的 50% 网点。传递函数的设定可以各个色版不同,也可以所有色版调用同一条曲线。如果采取这一补偿措施,那么在保存分色文件时要保存为 DCS 或 EPS 的格式,同时勾选"包含传递函数"选项,如图 6-13 所示。

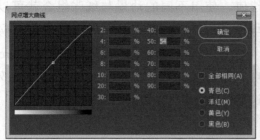

图 6-13 传递函数对话框

需要指出的是,Adobe 公司并不提倡用户在 Adobe Photoshop 软件中进行转移函数的设定,而是建议用户在照排机上输出时,采用照排机自有的标定程序来做这一步(参看 Adobe Systems Incorporated。PostScript Language Reference Manual)。

2. 印刷网点增大补偿

印刷的网点增大是指在印刷过程中,油墨被纸张吸收时因扩散而引起的网点增大。印刷的网点增大使得印刷出来的网点面积与实际所需要的网点面积存在差异,从而导致印刷实物色彩再现不准确。

不同的印刷机和不同的纸张组合会有不同的网点增大值。值得一提的是,对印刷的网点增大补偿并不能由测量值与理论值做减法得到补偿量,而是由网点增大曲线,根据反函数关系得到补偿值。以 50% 处的网点 A 为例,网点增大后得到的网点面积为 75%(点 C),但实际并不是以 75%-50%=25% 的网点面积来实现补偿,而是由点 B 向 Y 轴做垂线,与曲线的另一个交点 D 所对应的 X 轴上的点 E(30%),才是网点增大补偿后的网点面积。

印刷过程产生的网点增大一般在 Adobe Photoshop 软件中分色时进行补偿,可以对彩色图像的整体阶调或 CMYK 四个独立通道分别进行网点增大补偿设定;另外 Adobe Photoshop 软件还提供了对灰度图像(在分色时要转化为灰度图才有效)以及专色的网点增大补偿。

(1)利用 Adobe Photoshop 分色设置网点增大补偿

①在 AI 中制作带有不同网点面积率的色块测试文件;分别做出 CMYK 单色的网点色块以及四色叠印的色块。

②把制作的测试文件在既定的印刷条件下印刷后,用反射密度计分别测量各个色块的实际网点面积。

③在 Adobe Photoshop 中打开测试文件,选择"编辑"→"颜色设置"→"自定 CMYK"选项,在"网点增大"处选择"曲线",把测量得到的网点面积输入"网点增大曲线"对话框相对应的位置。

④Adobe Photoshop 内部会自动插值计算并生成针对该印刷条件下(油墨、纸张、印

刷压力等）的网点增大曲线，以后在该条件下印刷时，制作得到的该印刷网点增大曲线对当前的网点扩大有最好的补偿效果。

此外，还可以在"自定 CMYK"中选择"标准"，通过修改"油墨颜色"，就可以调用已经设置好的网点增大补偿值。使用不同的油墨与纸张组合就可以调用不同的网点增大值，如欧洲油墨标准与涂料纸、胶版纸、新闻纸搭配使用时网点增大分别是 9%（Eurostandard coated）、15%（Eurostandard uncoated）、30%（Eurostandard newsprint）。另外，如果制作了某印刷条下的 ICC 文件，那么除了自定网点增大外，还可以在"载入 CMYK"中调用 ICC 文件进行分色，也可以达到网点补偿的目的。

（2）制版流程印刷机补偿曲线

制版流程印刷机补偿曲线目的主要是找到印刷机的网点增大特性，根据目标网点增大数据生成补偿曲线。如 175 lpi 方圆网点在 50% 网点处的增大率一般控制在 15% 左右。具体操作方法请见前面相关章节。

3. 文字与图形的网点增大补偿

实际生产中，很多颜色是在图形类软件（如 Adobe Illustrator）和排版软件（如方正飞腾）中进行编辑的，但是这类软件都没有网点增大补偿功能，这就需要在设置颜色时考虑印刷网点增大后的颜色，专业的做法是在进行这类色彩设置时要有印刷色谱（色标）进行参考，如潘通（PANTONE）色标。

第七章 制版设备测试与异常处理

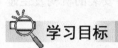
学习目标

1. 能调节设备的激光能量参数；
2. 能判定印版着墨不良问题；
3. 能判定并处理印版的质量问题。

第一节 CTP 印版输出与测试

CTP 印版检测是印版输出工作的一个重要环节，也是印版正常输出与保证质量的基础。在进行综合检测前的准备工作包括：

（1）用万用表检查，两组 AC220V 输入端内阻 (打开开关)，开关电源 24V、5V 的内阻情况，以确定电路中没有短路或断路现象；

（2）各驱动器的电源线颜色线序是否正确；

（3）光鼓码盘 0 位的确定，当光鼓前规转动在丝杆侧与光鼓轴心处于同一水平面时码盘的 LP 指示的 LED 灯亮，可用钢尺测量；

（4）两伺服电机的参数设置，参照工艺文件。

一、复位测试

主要对安装在光鼓圆周方向上的执行机构进行复位测试，该功能检查所有电机是否处于零位状态，在零位状态时，传感器 LED 灯亮后，各执行机构都在最高位，到各自的硬件限位约有 0.5 mm 的间隙。

检查部位如下：

（1）版头版尾装卸版电机传感器；

（2）装版压版辊电机传感器；

（3）摆辊抬升电机传感器；

（4）装版挡版电机传感器；

（5）左右平衡驱动电机传感器。

如程序命令不能运行完，会有相对应的提示，如等待某个电机或等待某个传感器，则需对相应的电机驱动器进行检查，驱动器的输入是否正确，输出是否正确；若是传感器，则要检查传感是否有效，安装是否正确。

执行复位测试的重要性是确保光鼓运转时，光鼓不会与沿光鼓周向安装的机构发生碰撞。从而保证设备运转的安全。

二、装版版头测试

装版版头测试主要对版头定位的准确性、版头机构控制的正确性进行检测。

三、卸版版头测试

卸版版头测试主要对光鼓卸版定位的准确性、卸版机构控制的正确性进行检测。

四、版尾测试

版尾测试主要对版尾定位的准确性、版尾机构控制的正确性进行检测。

五、平衡块位置与驱动测试

平衡块位置与驱动测试主要对平衡块位置的准确性，以及平衡块控制与调整的正确性进行检测。

六、进版压力、光鼓真空压力、放气、吸尘泵测试

进版压力、光鼓真空压力、放气、吸尘泵测试主要为进版压力、光鼓真空气压、放气、吸尘泵的测试与调整。

七、侧拉规机构测试

侧拉规机构测试主要对侧拉规的复位精度、对版材的侧边定位进行测试。

八、扫描聚焦镜头与物距镜头测试

扫描聚焦镜头与物距镜头测试主要对扫描平台上的聚焦镜头和物距镜头的控制与调整。

九、装版测试

装版调试主要调试版材如何正确装到光鼓上，以及如何判断装到鼓上的版材是否符合

设备工作要求。

1. 有无版检测传感器测试

在光鼓斜上方适当位置安装了一个反射传感器用于检测光鼓上是否有版材，在装版时，设备会检测光鼓上是否已有版，如检测有版，信息栏中会有提示信息，并要求先卸版。

2. 装版后检测

（1）调整侧拉规位置

版装到光鼓上后，用直尺测量版材边缘两边到光鼓边缘的距离是否一致。若不一致，调整侧拉规参数，确保版处于光鼓居中状态。

（2）确定版尾夹夹版的宽度

用铅笔在版尾边缘的两头和正中间画一条直线，将版从光鼓上卸下，用直尺测量画线到版边的距离，要求为 4～5mm，平行度 <0.2mm，如有不准确，则调整对应参数。

（3）版尾位置的精确定位

如果发现版尾位置逐渐减小，递减的平均值为 5，需将对应的参数加上递减的平均值。反之，如果出现递增的情况，则减去递增的平均值。

3. 版歪斜测试

执行装版测试，信息提示窗口中，[版尾斜度超标：××]，×× 值的变化应 <±4。

4. 标准版长调整

确定所用的版是标准的，且版长准确是标准的矩形。

5. 平衡块对应光鼓位置调整

用直尺测量，旋转光鼓，使平衡块检测杆中心的高度与设备的光鼓轴心高度一致。此时，不要让光鼓转动，并据此中心高，在版上用铅笔做好标记。如测量不便，可以用墙板撑板作为一个辅助基准。如平衡块在版头的外侧，可把标记做到光鼓上，并用卷尺测量到版头的长度，可分段测量。重复操作步骤，测量另一个平衡块并做好标记。如另一个平衡块在版尾外侧，可把标记做到光鼓上，并测量到版尾的距离。需要注意：平衡块调整不准确，光鼓旋转时，振动、声音都会较大。

十、激光聚焦、能量测试

1. 焦距调整

选择能进行焦距测试作业的 TIF 文件，如焦距测试图文件"Test_focus745X605_2400_90"；设置输出的版材上有多少条焦距条；根据设置的参数，输出焦距测试图作业。放置版时，把版头向上用放大镜观察焦距线，如果焦距线中有粗一点的线则表明物距偏小，应该将物距参数加大，反之则减小；找到相对应 dpi 焦距数值进行修改。

最佳焦距分布有一定的规律，最佳焦距呈现的焦距线是粗细均匀，呈现的网点是清晰完整的，最佳焦距的两侧网点是对称性的，离最佳焦距越远网点丢失的越多，直至看不到网点。最佳焦距的辨别方法如下：把版按版尾边靠近人的身体放在平台上，测试图横向分布着1%～99%的网点区域，用放大镜从右向左看1%的网点区域，在第 M 列中能观测到1%

的网点，对该列进行标记。随后会发现网点越来越清晰，然后又开始减少，如在第 N 列 1% 的网点消失，再对该列进行标记。此时找到两列的中间区域，应该就是最佳焦距。

2. 激光能量测试

激光能量功率测试时，需对比焦距测试步骤，选择能进行焦距测试作业的 TIF 文件进入作业模板设置界面，勾选"功率测试"，设置"最小功率"。每台机器都有其合适的功率范围，设置合适的起始功率；根据设置的参数，输出功率测试图作业；选择合适的冲洗条件进行显影。

根据以下几项内容选取最佳的能量：

（1）最佳的能量区域需没有底灰，可用丙酮或者无水乙醇滴到空白区域，空白区域没有任何变化为无底灰，如果出现扩散的蓝色圈，则为有底灰；

（2）最佳的能量区域网点还原要控制在 ±1% 以内，可以用网点测试仪进行检测；

（3）最佳能量区域 1% 的网点和 99% 的网点呈现完整清晰；

（4）根据以上三点选择的最佳能量值要在 CTP 的能量范围值内，并离上限的能量值要有一定的距离，如若接近上限或者超出范围，需要适度调整冲洗条件，直至能量值达到要求。

十一、版面图像位置调整

版面图像位置调整，首先输出一张平网，在作业模板参数设置的输出位置，上下边空为 0，勾选横向、纵向居中，检查以下关键点。

（1）网线中不能有任何干扰条纹；检查各个区域的网点增大率是否一致；若版面上有暗点，这种情况一般因为光鼓有脏点，导致版装入光鼓后局部有凸起，凸起部分会因为离焦产生暗点；

（2）测量版头图像到版头版边缘的距离，如果与设计理论值不符，则调整对应参数，版头边缘与曝光点间的距离，十个参数为 1mm，可以输入小数点；

（3）测量图像边缘到版边缘（光鼓横向）的距离，如果与设计理论值不符，则调整对应参数，记录头曝光点到作业输出点距离，十个参数为 1mm，可以输入小数点。

十二、直线电机系统的维护

针对直线电机光栅定期维护，可以提高激光副扫描机构的运行精度和稳定性，保证网点质量；直线伺服系统光栅清洁时严禁使用酒精、丙酮等液体，只可用干棉布清洁。

十三、自动加油系统维护

检查设备副扫描机构润滑状态及导轨储油量，副扫描机构是自动润滑，油定期加满后，可实现免维护，有助于提高激光副扫描机构的运行精度和稳定性。

第二节 异常情况分析及处理

一、图像质量问题及处理方案

输出印版时，若版面上出现质量问题，则引起此问题的原因可能是如下几种。

（1）计算机：计算机的兼容性有问题、计算机存在病毒；

（2）制版机：焦距偏离、功率不对，输送过程中或光鼓上有毛刺，激光不受控，丝杆导轨缺油，机构调整不当；

（3）冲版机：显影温度、速度设置不合理，保养清洁有问题使设备不能正常工作，各辊轴、毛刷压力调整不当，补充液配比不合理；

（4）版材：版材封孔不好、感光胶涂布不均匀，版材的平整度超标，版材的边缘不规则，不适合制版机的版材。

二、其他异常情况及处理方式

硬件系统在工作过程中，参照综合调试流程，根据报错信息，进行相应的调整。

1. 激光锁不定问题及处理方法

激光锁不定的原因如下：

（1）出光未对准能量检测传感器探头中心；

（2）镜头、密排、能量检测传感器表面有灰尘；

（3）连线的线头接触不良；

（4）激光驱动板有问题；

（5）激光箱的恒温控制有问题；

（6）光纤与激光器耦合有问题；

（7）光纤损坏；

（8）激光器损坏。

2. 确定激光箱的恒温控制系统是否符合工艺要求

根据程序的信息提示栏中的信息，锁光时，有温度提示，温度要求控制在 23.5～26.5℃范围内。

恒温激光箱的装配要求如下：

（1）打开软件，执行读取功率命令，根据反馈的数值判断激光箱的温度，要求温度为（25.5±0.5）℃，如温度过高，需检查激光箱恒温箱部分散热风扇工作是否正常，恒温驱动板工作是否正常，制冷片工作是否正常；

(2)检查散热风机工作状态,观察恒温控制箱内散热风机的工作状态,是否有故障风机的存在,如有,需更换散热风扇。

3. 确定出光点位于光电池的中心

(1)平台复位。

(2)开启中间的某路激光,观察光斑位置,要求光斑位于光电池的中(目测或轻轻转动光电池位置,当通过软件读取激光功率的值最大时即为中心位置),如图7-1所示。如有问题,需调整光电池的固定位置,使其满足该要求。

(3)更换标准光电池,锁定光功率,如果光电池有问题,需更换。

图7-1 测试出光点

4. 灰尘污染造成激光功率下降

灰尘污染是造成激光功率下降的最常见的问题,UV机型最易受污染的位置包括镜头表面,密排头出光端污染。因镜头作为外露的元件,非常容易受灰尘及其他杂物的污染,所以出现激光功率下降,首先要检查激光镜头处是否存在污染。

擦镜头前,先用激光软件检测激光能量值,然后通过命令操作中的精度调整指令,使镜头向后移动,然后关闭电源,手动将扫描平台移动至右限位。用擦镜纸或不掉毛的棉签蘸少许清洁剂(10%的乙醚和90%的无水酒精混合)或直接使用无水酒精,擦拭时,沿镜头的一个方向擦拭,光电池出光点也一并清洁。清洁完毕后,开机完成精度调整指令,使镜头复原,然后用激光软件检测清洁后的激光能量值,若发现能量值低于擦拭前,重复上述清洁工作(见图7-2)。

图7-2 擦镜头

密排头污染也会造成激光功率下降,密排头污染的清洁过程如下。

(1)使用指令,密排筒座移动到复位点,如图 7-3 所示;

图 7-3　密排筒座

(2)关机,去除上罩盖和后上罩盖;

(3)将卸版架的电源线和信号线断开,松开紧固螺丝后,移出卸版架;

(4)去除扫描平台罩壳,用铅笔在镜头筒与密排筒座上做连线标记,确保复原时位置不发生变化;

(5)拧松镜筒保护套的十字螺丝,然后旋转镜头筒使之脱离密排筒座,如图 7-4 所示,然后对密排表面进行清洁,如图 7-5 所示;

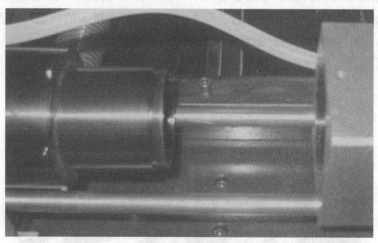

图 7-4　镜筒分离

图 7-5 密排头清洁

（6）用无纺布蘸少许无水乙醇，沿一个方向擦拭密排头，擦拭一次后更换无纺布擦拭部位，直到看不到脏点；

（7）开机，然后用激光软件检测激光能量值，确保能量值有所提升后，完成密排头清洁；

（8）复原镜头筒，装回平台罩壳，装回出版架，恢复机器外罩壳。

5. 某路或某几路激光功率下降

某路（假设为 X 路）激光未锁定，可能原因是激光驱动板、连接线头部松动、激光器与光纤耦合不良、光纤损坏、激光器损坏，可用互换排除法确定问题原因。

（1）确定驱动板

把 X 路连接激光器的插头与相邻的某路（锁光正确）互换，锁光，如果功率降低跟着 X 路，那就可确定驱动板卡没问题。反之，需检查驱动板卡的问题，检查驱动板卡的联线是否可靠，若还是有问题，则需更换驱动板卡（见图7-6）。

图 7-6 检查驱动板卡

(2）确定激光器与光纤耦合

①用一字螺丝刀，插在陶瓷部金属槽中，慢慢推动光纤旋转一个角度，如60°；

②测试功率，如果功率恢复正常，再用螺丝刀轻微摆动光纤，如果输出功率稳定，没有变化，那就是耦合有问题。

(3）确定是光纤的问题，还是激光器的问题

①采用光纤互换的方法，确定是光纤问题还是激光器问题，如图7-7所示；

②用镊子，轻轻地将 X 路的光纤末端陶瓷芯推出激光器，同样操作 X±N 路（X±N 路是正常的）；

③清洁光纤头，用专用的光纤头清洁器清洁光纤头，或用无尘纸清洁光纤头，在显微镜上，光纤芯径上看不到任何灰尘；

④把 X、X±N 路互换插入；

⑤锁定激光，如果还是 X 路功率低，则可确定光纤密排有问题，需联系科雷客服部；如果是 X±N 低，则说明功率跟着激光器走，激光器异常需更换激光器。

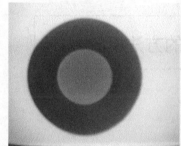  

图 7-7　确定光纤与激光器的问题

(4）更换激光器

操作时，需戴防静电装置保护激光器。

①拔下联接激光器的插头，用内六角扳手，拧开固定激光器的压板螺钉，取出激光器，并放好；

②把新的激光器放入原位，激光器底面有导热硅脂，用压板固定激光器，拧紧螺钉，拧紧力适度；

③把激光器的插头，插入对应的位置；

④清洁光纤头，用显微镜观察，无任何涨点后，插入激光器的内孔；

⑤通电，检测，锁激光，功率确定达到标准；

⑥轻微转动光纤，功率不变；

⑦按（4）的⑤、⑥步操作，点胶固定，垫好；

⑧复位激光箱；

⑨做好更换的记录。

（5）确定某路激光不受控的方法

运行程序，在功率补偿配置窗口中，鼠标双击 0 路对应的数值，如改为 50，意味着把 0 路的激光功率，改成为原锁光功率的 50%，在成像时，由于此路功率不够，对应此路激光的线就会变粗（阳图版），以此路为基准，就可找出有问题的光路，单击左上角">"符号按钮，则可以把所有百分值一起配置到对应数字光路中，锁定功率，检查是否已按要求锁定。

发送焦距测试图到 CTP，对照焦距测试条的线条部分，就可以检查，把版头放在前面，若非 0 路的有某激光线有问题，可从错开的焦距线中比较，0 路的线比较粗，如果 0 路在下侧，左侧相邻的即为 2 路，左上侧为 1 路，依此类推，找出不受控的那路激光，查找此路激光的各连接线头是否有松动，如问题仍存在，换激光驱动板。注意：用显微镜观察时，类似倒像。

第三节　判定印版输出质量的方法

一、查阅工作指令、工艺要求

CTP 出版工在接到印刷工艺单以后，需按印刷工艺单上叼口要求对文件咬口进行调整后出版。出版前应在计算机预览中检查咬口无误后再输出。

印版输出前 CTP 出版工必须在输出列表中仔细核对需要输出的"产品名称"和胶印工艺单上的"拼版对象"名称是否一致，检验一致后在胶印工艺单"拼版对象"后签上检查人的名字作为检验的凭证。未经过此道检验的文件不允许输出。

作业输出设定后，放置印版并按下绿色按钮（或 LOAD 键），印版曝光完成后自动进入冲版机，将输出印版摆放于"待检"架子上等待检验。

1. 自检

印版完成后直接制版机输出人员对印版进行质量自检，检验版面位置，叼口方向，叼口大小；检验印版信号条，网点还原正常，版面无污迹；并在检验合格后将印版放置于"合格"架子上。

2. CTP 版材等版情况及特殊烤版要求的确认

凡使用 CTP 版材的产品，一律不需要做上车版备用印版。使用 CTP 版材的产品大于 8 万印的应烤版。

3. CTP 质量专检

按照印刷施工单或标准样张为检验标准；印版检验员根据当日所需检验的印版，逐一进行检验。

检查一套印版色数是否俱全；版面角线下方处"标的"的内容：包括编号、产品名称、制作人、规格、色数、日期、色标是否齐全。尤其是"产品名称"，必须和胶印工艺单上的"拼版对象"名称一致，此检查项目为判断印版是否合格的一个必须条件。

（1）检验印版是否与签样（承印标准样张）的规格、彩图及文字一致；

（2）检验整套印版咬口是否符合印刷施工单上的胶印工艺要求；

（3）检验每一张印版尺寸是否正确，是否有脏点等缺陷；

（4）检验完毕后做好记录；

（5）每一张印版检验合格后，在拖梢处用白粉笔标注好产品名称、色别、日期；

（6）凡检查出有问题的印版，同时反馈前道工序进行返工、返修或重新制作，将不合格印版做好状态标识和隔离工作，如需报废销毁的应做好销毁记录；

（7）印版发放应按领版申请的订单号、产品名称、色别、车号等内容，对印版进行准确发放，并做好印版出库记录；

（8）报废的印版，必须按一般废弃物的处置规定，集中堆放，统一处置；

（9）印版制作叨口，按印刷工艺单；

（10）印版存放条件应控制在温度17～26℃，湿度35%～70%RH，原包装保质期一年。

二、成像焦距调整

1. 初调

方法及要求：

（1）"工程师操作"→"修改设备参数"→双击"19"号参数的数值（如表7-1所示），19号对应的值改为1200→"确定"；

表7-1 热敏/UV对应的19号初始值

激光器类型	19号参数对应的初始值
405nm 激光器	1200
830nm 激光器	1500

（2）在"命令功能操作"中选择"模板调整"，然后选择"命令装版"，装一张没有曝过光的版材；

（3）进入LaserAdjust 4.10.exe程序，打开其中一路激光为常开状态；

（4）"命令功能操作"→"平台移动"→"发送命令"，将记录头移到激光可以找到光鼓有版材的范围（移动时准备一个挡片等记录头到版材范围时，将2204号传感器挡一下，接着挡一下2203号传感器，记录头才会停在指定的位置）；

（5）松开镜头座上的锁紧镜头螺钉（两颗M3内六角），前后移动镜头并慢速转动光鼓，观察激光点到最小现象，并且会在版材上烧灼出一条很细的白线，此时为初调的最佳焦距，紧固两颗螺钉；

（6）检查该结构上的所有螺钉的紧固性；

（7）关闭激光，退出 LaserAdjust 4.10.exe 程序；

（8）"命令功能操作"→"命令卸版"→"发送命令"。

2. 焦距、物距细调

进入软件，选择能进行焦距测试作业的 TIF 文件，如提供的焦距测试图文件 "Test focus 745×605_2400_90"，如图 7-8 所示。

单击"操作"→"打开输出文件"到输出作业队列，鼠标右击此作业，在下拉菜单中选择"参数设置"，进入作业模板设置界面，勾选"焦距测试"，如图 7-9 所示。

设置"聚焦测试"（建议数值为 20），其单位"次"表示输出的版材上的焦距条数，和"聚焦补偿"（建议数值为 -40）。

根据设置的参数，输出焦距测试图作业。放置版时，把版头向上，如图 7-10 所示。

在版子上找到合适的焦距，方法是用肉眼观察焦距测试块中的焦距线都应全部均匀排列，没有条纹，如图 7-8 所示。或用放大镜从右往左，观察 1% ～ 2% 的网点从开始有到没有都分别做上记号，然后从两边往中间数，取中间的一个，再观察焦距测试块中的焦距线都应全部均匀排列，如果有明显的线细一点或粗一点都是不正常的，1×1（像素 × 像素）的网点里排列应很均匀不应出现明显的深浅不均匀的现象，2×2（像素 × 像素）的网点应绝对均匀不得有任何干扰纹路，如图 7-11 所示。

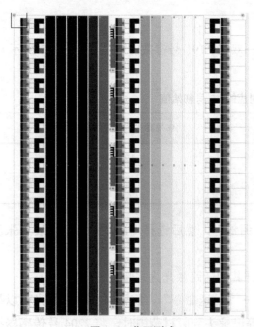

图 7-8 焦距测试

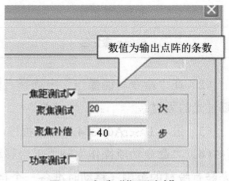

图 7-9 勾选"焦距测试"

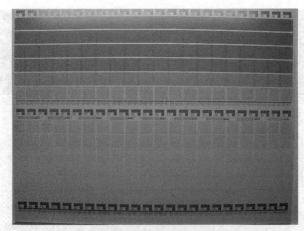

图 7-10 输出焦距测试

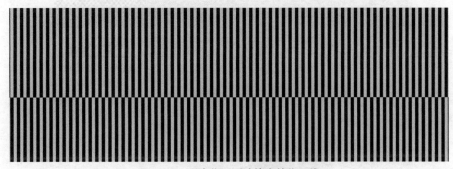

图 7-11 观察焦距测试块中的焦距线

例如：把版按版尾边靠近人的身体放在平台或桌面上，用放大镜从右向左看，第 M 个出现 1% 的网点，过了 N 个后，1% 的网点消失，N/2 的那一块，应该焦距最清晰。那么，焦距的最佳位置位于 M+N/2。

注：M 是从 0 开始数的，如 N 是奇数值，可看 N/2 中左右两块中的小测试块中的线条，比较确认相应的焦距变化值即为（M+N/2）×4。

如此，焦距最佳值在从右边数过去的第 12 块上（从 0 开始计数），则焦距的变化值为 12×5=60。

$$f=F+(5·X)=-50+(5×12)=10$$

用放大镜观察焦距线（见图 7-12），如果焦距线中有粗一点的线则表明物距偏小，应该将物距参数加大，反之则减小。

同时，还要观察实地部的边缘，如果边缘不齐，还需要调整密排的角度（密排斜排方式），调角度的过程中会影响到物距、焦距的大小，反过来物距、焦距的大小也会影响到角度的变化，调整的时候应注意这个问题。

在 10%～99% 的角度为 90°的平网内不能出现任何干扰条纹。

进入"工程师操作"→"修改设备参数"（见图 7-13），找到相对应 dpi 的焦距数值进行修改。

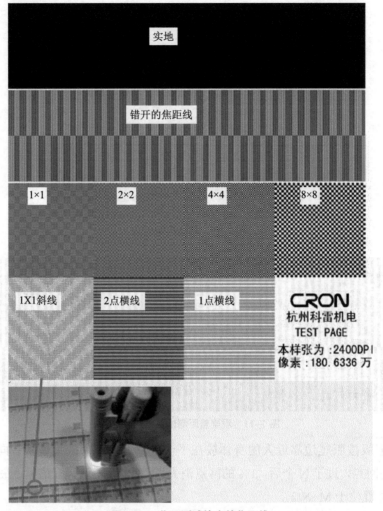

图 7-12　焦距测试块中的焦距线

如此次为 2400 dpi 焦距测试，焦距调整的值为 10，原 19 号参数值为 1200，加入后，更改为 1210，单击"保存设备控制参数"。

检查版尾压痕是否锐利没有虚影，主要验证版尾压条是否整条压实了版边。

三、光功率调整

宏观上看，图 7-13 的 1×1（像素 × 像素）网点部分，有没有条纹，用放大镜看，有没有深浅变化，如果有，则需对光功率进行补偿，如图 7-14 所示。

设置"最小功率"，每台机器都有其合适的功率范围，设置合适的起始功率，根据设置的参数，输出功率测试图作业（如图 7-15 所示）。

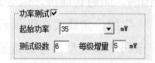

图 7-13 光功率不均匀　　　　　　　图 7-14 勾选"功率测试"

确保冲洗条件正确后，在输出的版材中挑选网点效果最好的焦距条（版头向上时，一般在 1% 网点处右边的网点效果比较好，而在 99% 网点处左边的网点效果比较好，选择时选择网点都比较均匀的网点焦距条），并以版头向上的位置，从右向左第二个焦距条开始数。第 X 条为最佳焦距条，当前激光功率为 P，每级增量为 p1 补偿后合适的激光功率为 p，则最合适的激光功率为

$$p = P + (p1 \times X) \tag{7-1}$$

用专用仪器测量：网点的增大率，一般测出的结果都比理论小 3 个百分点左右为合适；空白的区域的底灰标准在 0.005 以内（可以滴一滴丙酮在空白处，散开后应不能看到明显的色圈）。

取消"功率测试"，在激光功率后的下拉框选择与 p 值相等的激光功率，单击"确定"保存此参数，并对模板及待输出作业的激光功率进行更改。

图 7-15 仪器测试与分析

第四节 印前处理及制版与印刷质量的关系

一、感光片和印版质量要求

（1）胶片和 CTP 版套印准确，误差≤0.05mm；实地平实光洁，无明显墨杠、不花，网点均匀清晰。

（2）版面整洁、色标齐整，信号条各色版不重叠，紧贴出血线，高 5mm，所有颜色做实地，铺满。如遇到混拼，信号条的颜色与产品要相对应。

（3）十字线、角线要在纸张范围内出现，如不能出现要适当延长。

（4）条码清晰完整，符合国家标准。输出软片的请叠色扫条码，必须≥B 级，输出 CTP 的，条码部分应先出软片，按软片检验标准执行。

（5）每一拼联均与批样要求一致，版面排列，尺寸符合图纸或划样要求。每一联按常规要求做好专色与专色，专色与四色之间的陷印，有特殊要求的请按备注或批样要求制作。

（6）标的齐全，位置正确，标的最上端离净角线 25mm。

（7）如遇特殊咬口的产品，请按备注或批样要求制作。

二、图像扫描的质量

（1）阶调齐全、定标准确，清晰度高。

（2）扫描图像要接近原稿、高于原稿。

（3）图文组版要考虑后工序。

（4）输出质量应符合字体完整、图片链接准确、陷印正确。

（5）印前数字流程版本尽可能高、能兼容最新版本的 PDF 文件。

（6）数码打样质量应做好打印机标准化工作，打印时选择正确的特性化文件。

第八章　检查数字打样样张

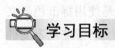

学习目标

1. 能使用测量仪器测量样张的各项技术参数；
2. 能使用测控条检验样张的质量；
3. 能提出并实施打样机的周、月保养计划。

第一节　样张质量检测与控制方法

　　一般来说，出版商或其他客户只有在得到印刷厂提供的打样样张，签字确认后，印刷厂才正式开始执行印刷客户所需的印刷品。在这个过程中，包装设计师，印刷厂或打样公司根据制版公司提供的胶片或电子文件，制作印刷样品的过程称为打样。客户对印刷样品的版式设计、印刷质量进行检查并签字确认样品可以作为印刷的根据，这个过程称为签样。

　　数码打样是以印刷品颜色的呈色范围和与印刷内容相同的 RIP 数据为基础，采用数字打样设备来再现印刷色彩，并能根据用户的实际印刷状况来制作打样样张的过程。理论上来说，数码打样设备色域大于印刷色域，就可以通过软件控制，进行输出设备间的色彩空间转换，使数码打样效果模拟印刷输出设备的色彩效果，即数码打印机的色彩管理过程拼版，实现数码样张代替传统样张签样。数字打样伴随 CTP 直接制版技术的发展而发展起来，是数字式工作流程中不可缺少的一个环节。

　　目前数码打样系统由数码打样输出设备和数码打样控制软件两个部分构成。其中数码打样输出设备是指任何能以数字方式输出的彩色打印机，如彩色喷墨打印机、激光打印机等。但目前能满足出版印刷要求的打印速度、幅面、加网方式和产品质量的多为大幅面彩色喷墨打印机；数码打样软件有 EFI 数码打样软件系统、网屏数码打样系统、方正数码打样系统，等等。

一、数码打样中的控制过程

1. 设备校正——打印机线性化

普通彩色喷墨打印机的线性都存在一定的问题，其表现为大于 90% 的暗色无法区分阶调的变化而出现并级现象，而且各打印原色的线性也不相同。如果用这样的打印机输出的标准色表制作样张输出设备的特征文件洗涤用品包装，则会使输出设备的特征文件反映的设备特性产生误差。因此，打印机线性化是实施数码打样色彩管理的第一步。注意：更换纸张与墨水等耗材后或者人为调整打印机后，必须重新做线性化。

2. 特性文件的制作

打印机特征化是进行色彩管理的一个十分重要的环节。其基本过程是使用标准色表文件如 IT8.7/3 或 ECI2002 等，通过数码打样软件和彩色打印机，打印出一张标准色表文件的数码打样样品。通过分光光度计和专用软件进行测试和计算，最终获得一个反映彩色打印机和打印纸张特性的特征文件（property file）。

3. 工作流程中色彩管理的设置

打印机线性化及打印机的特性文件创建完成后，为了实现数码样张与印刷样张的匹配，需在数码打样工作流程中进行设置，让数码打样效果模拟印刷输出设备的色彩效果。

二、影响数码样张的因素

数字打样是指以数字出版印刷系统为基础，在印刷生产过程中按照印刷生产标准与规范处理好页面图文信息，直接输出彩色样稿的新型打样技术。数字打样终极目标是模拟最终的印刷效果，基本思路是先建立一种比较正常的、容易实现的印刷效果，作为数字打样的理想目标，再根据这种效果调整数字打样，保证其比较接近理想目标。

1. 打印机

（1）色域要求

为了适应未来高保真印刷和个性化印刷的需求，数码输出设备要有能够表现印刷效果的能力，最好能够采用 8 色或 9 色打印，从而实现更宽广的色域表现能力。

（2）打印精度要求

喷墨打印机打印头工作情况的好坏将直接影响数字打样的输出效果。打印头能够达到的打印精度决定数字打样的输出精度，低分辨率的打印机无法满足数字打样的需求。

2. 墨水

打印墨水对打样色彩的还原起到决定性作用。颜料墨水有利于印品的保存，印品不易褪色，其墨水原色同印刷油墨更加接近，但光源环境对样张色彩影响更加明显。

3. 纸张

数字打样所用纸张一般为仿涂料打印纸，一方面，它同印刷用涂料纸具有相似的色彩表现力，更易达到同印刷色彩一致的效果；另一方面，其表面具有适合打印墨水的涂层，涂层的好坏会影响样张在色彩和精度方面的表现，同时打样纸张的吸墨性和挺度也会影响打样质量。

三、样张质量测试方法

样张质量进行检测与控制的方法主要有主观目测法与色度检测法。

1. 主观目测法

采用人眼观察样张，根据经验判断样张的质量。该方法受人为及外界因素影响较大，不能全面反映样张的质量特性，但能判断样张最终的质量好坏。

2. 色度检测法

色度测量是从样张画面上直接测量测试点，得到颜色数据，通过计算检测样张和标准样张上相应点的色差值，根据色差值大小来判断检测样张与标准样张上的颜色是否一致，以判断该位置颜色的再现是否正确。其优点是对颜色的判断结果与人眼的一致，准确性高，从而避免了人眼对颜色进行判断时的主观因素及疲劳等造成的误差影响。但当色差值大于规定的色差上限时，不能反映印刷过程中控制因素的改变量。

第二节 数码打样测控条

一、Fogra 数字测控条

评价和认证数字打样的质量，需要一个数字打样认证软件。色彩管理认证软件是用数据来确认数字打样的色彩管理的正确与否。这些软件是由国外的一些公司开发的，目的是定期对数字打样的色彩管理进行数据上的认证，可改善主要用目测的方法确认数字打样色彩的缺陷。对不认可的数字打样系统及时进行校正，以保证数字打样的色彩准确性，从而使色彩稳定。

数字打样认证软件一般是选择国际通用的印刷标准进行认证，如 ISO12647-2、SWOP、GRACOL、FOGRA，也有企业自定义的印刷标准，并采用由这些标准定义的彩色测控条，用数字打样软件打印出来，一般可打印在作业的四边。

在数字打样的色彩认证中，采用 FOGRA 39 V3 标准的测控条，共 72 个色块，作为数字打样软件的测试条，如图 8-1 所示。

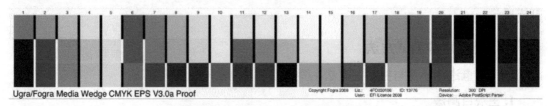

图 8-1　FOGRA 39 V3 标准的测控条

数字印刷测控条由 3 个模块组成，其中模块 1 和模块 2 用于监视印刷复制过程，模块 3 则用来监视曝光调整，各测量控制色块的尺寸大约为 6 mm×6 mm（有时可能小于或大于这一数值）。由于模块 3 包含的控制块主要用于监视数字印刷的曝光记录过程，故设计得与 UGRA/FOGRA 用于检验和控制软片输出的 PostScript 测控条对应。

1. 模块 1

包含以下 8 个实地色块：青、品红、黄和黑色实地色块各 1 个，"青＋品红""青＋黄""品红＋黄"实地色块 3 个，"青＋品红＋黄"实地色块 1 个。这些控制色块用于控制数字印刷油墨的可接受性能以及三种减色主色的叠加印刷效果。

2. 模块 2

（1）颜色平衡控制色块：该色块为规定的灰色调数值，与胶片输出有关。它实际上包含两个色块，其中右色块为 80% 的黑色，用于控制网目调加网效果；左色块由 75% 的青、62% 的品红和 60% 的黄组成，目的是与 80% 的黑色色块比较。印刷时若灰平衡控制不好，则该色块将呈现出彩色成分。

（2）实地区域：实地区域包含 4 个实地色块，按黑、青、品红和黄次序排列，每隔 4.8 mm 放置一个色块。第一个实地色块（黑色块）紧靠颜色平衡控制色块，它的四个角上压印了黄色，用于检查印刷色序，即黄色先于黑色印刷还是黑色先于黄色印刷。

（3）D 控制块：D 为 Direction，因此 D 控制乃是指方向控制之意，即检验采用特定的复制技术、复制设备和承印材料组合在不同方向加网的敏感程度。

D 控制块分为四组，青、品红、黄和黑色各一组，每一组中均包含 3 个色块，其总尺寸为 6 mm×4 mm。在组成数字印刷测控条时，通常按黑、青、品红、黄的次序排列，位置在实地色块后。3 个色块均采用线形网点加网，加网角度从左到右依次为 0°、45°、90°，每个色块采用的加网线数均为每厘米 48 线，阶调值为 60%（60% 黑）。之所以采用 60% 阶调值而不采用中间调值（50%）的主要理由是，输出后的色块比中间调略暗，可以更清楚地识别加网工艺的方向敏感性。

理论上，当采用相同的加网线数和网点形状时，则这 3 个色块应该有相同的密度值。如果实际测量出来的 3 个密度值有较大差异，则说明用户使用的复制技术、复制设备和承印材料组合在某个加网方向上太敏感。

（4）网目调控制块 40% 和 80%：该控制块同样有青、品红、黄和黑 4 组，每一组控制块由 40% 和 80% 两个色块组成，采用 150 lpi 加网。这一数字与大多数商业印刷品采用的记录精度是吻合的。两个网目调控制色块与中间调网点百分比呈不对称分布，代表了比中间调略淡（接近中间调）和接近实地的网点百分比。不同的数字印刷工艺采用不同的加网复制技术，会得到不同的输出效果。因此，这两个控制块可用来评估特定数字印刷加网技术的表现能力与行为特性，衡量加网技术能否获得需要的记录效果。在形成测控条组合时，按黑、青、品红和黄的次序排列，位置在 D 控制块后。

3. 模块 3

该模块包括 15 个不同程度的灰色块，每个色块的尺寸相同（6 mm×10 mm），均采用黑色油墨印刷。15 个色块组成 5 列，每一列均包含 3 个色块，但采用了不同的网点结构。上述色块的油墨覆盖率分别为 25%、50% 和 75%，其中最左面一列为 25%，第二、三、四列油墨覆盖率为 50%，第五列为 75%。

只用黑色油墨印刷这些色块的原因很简单，是为了节省测控条占用页面的空间。控制块的第一行总是用输出设备可以达到的最高记录分辨率复制，第二行色块的记录分辨率是第一行的二分之一，第三行是第一行的三分之一。由此，从第二行和第三行色块可看到较大的网点结构。控制块的第二、三、四列均为 50% 黑色，第二列命名为 50cb（Checker Board），它们均是格子状图案；第三列包含水平线；第四列则包含垂直线。

理论上，模块 3 被印刷出来后，每一列色块的阶调值应该是相同的，不同的仅是记录分辨率；在行方向上，每一行中间 3 个色块复制到纸张上后也应该具有相同的阶调值。因此，如果每一行中间 3 个色块的阶调存在差别，则这种差别一定与复制方法有关，导致差别产生的原因可从网线角度方向上找。输出时应该将记录设备调整到使行方向的阶调差别最小。列方向上色块的阶调值不同时，反映了加网线数对复制效果的影响。

4. 模块的组接

数字印刷测控条的模块 3 单独使用，模块 1 与模块 2 的组接原则上是自由的，但为了排列得更有规则，可采用如下次序：先安排平衡控制块；接下来是黑、青、品红、黄实地块，再加黑色 D 控制块和黑色网目调加网控制块；后面是黑、青、品红和黄色实地块，再加青色 D 控制块和青色网目调加网控制块；再后面是黑、青、品红和黄色实地块，再加品红色 D 控制块和品红网目调加网控制块；最后是黑、青、品红和黄色实地块，再加黄色 D 控制块和网目调加网控制块。

二、数码打样国际标准 ISO 12647-7:2016

2016 年 11 月 15 日，数码打样国际标准 ISO 12647-7:2016 第 3 次修订版正式发布。2016 年 12 月，Fogra 发布公告提醒所有 Partner 新标准的更新内容及 PSO 认证即将更新为新标准。2017 年 2 月，CGS ORIS 推出 Certified Web V2.0.8 新版本，支持该项新标准。2017 年 4 月，Fogra 印前技术部经过一段时间的测试，发布了 Fogra Extra 36，向业界书面解释该标准。众所周知，ISO 12647-7:2016 是数码打样的国际标准，如图 8-2 所示。

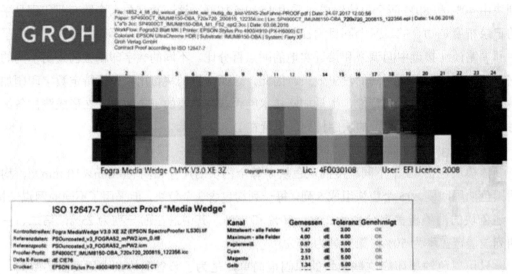

图 8-2 ISO 12647-7 数码打样的国际标准测试文件

一份完全符合 ISO 12647-7:2016 的数码打样，需要 MediaWedge 测试色块符合公差要求，还有承印物的亮度、荧光剂含量、耐磨性、耐光性等都有相应的要求。2013 年 ISO 12647-7:2016 发布后不久，关于三类通用的纸张白度需要亮度值 L 大于等于 95 的要求已经显得不理想。虽然它对于 Fogra 39 还比较适合，但对于其他的打样条件，如新闻纸，显然就不合适了。

因此新的 ISO 12647-7:2016 标准提出以下几个要求。

1. 光泽度（GLOSSY）

首先将材料分为亚光、半亚光、亮光三种类型，要求打样材料与印刷材料进行色彩匹配时，应该是三者中的同一类型。采用亚光打样纸作为光粉纸印刷跟色稿的做法，或者采用亮光打样纸作为哑粉纸印刷跟色稿的做法都是不合适的，如图 8-3 所示。

2. 纸白模拟（Paper Simulation）

没有图文地方的打样纸色彩应该允许模拟印刷纸张的色彩（即绝对比色），且色差 $\triangle E_{2000}$ 应该小于等于 3.0。而为了保证纸白模拟的准确性，打样纸的亮度值（L 值）显然应该要高于印刷用纸的亮度值（L 值），如图 8-4 所示。

Classification	75° ("TAPPI gloss")	60° (ISO 2813)
Glossy	> 60	> 20
Semimatte	20～60	5～20
Matte	< 20	< 5

图 8-3 光泽度标准

Classification	Description of OBA
0 ≤ ΔB ≤ 1	Free
1 < ΔB < 4	Faint
4 ≤ ΔB < 8	Low
8 ≤ ΔB < 14	Moderate
ΔB ≥ 14	High

图 8-4 纸张白度模拟

3. 荧光剂（OBA）

打样纸的荧光剂含量等级应该与印刷用纸属于同一等级。根据 ISO 15397:2014，荧光剂分为四个等级：微量、少量、中等、大量。实际上还有不含 OBA（荧光增白剂）的情况，因此通常为五类。这与 ISO 12647-2:2013 中的分类是一致的，也体现了数码模拟印刷的原则。

在 ISO 12647-7:2013 标准中没有明确的老化耐久性测试的要求，而 ISO 12647-7:2016 新标准中做了明确的规定。该测试需要四份相同测试样张，其中包含实地和网点的 CMYK 及 RGB 叠印色。

并且规定了四种模拟测试环境，以及在这种环境中处理前后的色差要求。①室温环境：25℃，25% 的相对湿度，24 小时；②高温湿环境：40℃，80% 的相对湿度，24 小时；③干燥环境：40℃，10% 的相对湿度，一周；④低曝光环境：依据 ISO 12040 规定的至少 3 个步骤。

4. 处理前后的色差要求

纸白色差ΔE_{2000}小于 2.5，其他色块的色差ΔE_{2000}最大值小于 2.0；如果是亚光材料或其他非常粗糙的材料，色差ΔE_{2000}可以放宽到 4.0。

控制色块色差是一个重要的变化。色差公式采用与视觉评估更加一致的ΔE_{2000}而不再是ΔE_{76}，由于ΔE_{2000}（*00）和ΔE_{76}（*ab）这两个公式不能互相转换，因此新版标准定义了新的容差要求。

两个色差公式计算方法不同，因此符合旧标准的数码稿未必符合新标准，而符合新标准的数码稿也未必能达到旧标准的要求。

图 8-5 为 ISO 12647-7 新、旧标准的参数要求对比，包括纸白模拟、所有色块、三色灰、一次色（CMYK）ΔH、一次色（CMYK）ΔE。

ISO 12647-7	Paper simulation
old (2007)	$\Delta E^*_{ab} \leqslant 3$
new (2016)	$\Delta E_{00} \leqslant 3{,}0$
ISO 12647-7	**All patches**
old (2007)	Maximum $\Delta E^*_{ab} \leqslant 6$ Average $\Delta E^*_{ab} \leqslant 3$
new (2016)	Maximum $\Delta E_{00} \leqslant 5{,}0$ Average $\Delta E_{00} \leqslant 2{,}5$
ISO 12647-7	**Composed grey**
old (2007)	Average $\lvert\Delta H\rvert \leqslant 1{,}5$
new (2016)	Maximum $\Delta C_h \leqslant 3{,}5$ Average $\Delta C_h \leqslant 2{,}0$
ISO 12647-7	**Primary colours**
old (2007)	Average $\lvert\Delta H\rvert \leqslant 2{,}5$
new (2016)	Maximum CMY $\lvert\Delta H\rvert \leqslant 2{,}5$
ISO 12647-7	**Primary colours**
old (2007)	Maximum $\Delta E^*_{ab} \leqslant 5$
new (2016)	Maximum $\Delta E_{00} \leqslant 3{,}0$

图 8-5　ISO 12647-7 新、旧标准的参数要求对比

为了评估新、旧标准的效果及影响，Fogra 对 116 份数码打样分别做了测试与分析，结果显示：ΔE_{76}色差公式对于比较饱和的色块，数据结果与视觉评估结果有比较大的差异，而新的ΔE_{00}则与视觉评估比较一致。

5. 专色评估（CxF）

这是打样流程标准第一次定义专色评估的要求。专色在包装应用中非常普遍，特别是在应对客户对颜色一致性的高期望中，往往只有采用专色才能达到客户要求。

但目前的数码打样系统，一般采用喷墨打样机，且不允许添加自定义的专色墨水，色域始终有限，因此这仅适用于需要评估的专色在打样机可呈现的色域范围内的情况。

在这种情况下，ISO 12647-7:2016 新标准规定参考色样与打样最大色差 $\triangle E_{00}$ 在 2.5 以内，而参考色样则由客户选择或提供，标准推荐使用 CxF/X-4 数据。

基于 CxF/X-4 的专色定义是比较理想的，它应该包含专色的光谱数据、垫白垫黑的数据、油墨透明度等，而这些都可以改善在客户与印刷厂之间的色彩沟通。专色的标准化还为时尚早，因此强烈建议要事前与客户协商好，特别是专色在打样色域之外的情况。定义专色的 Lab 值是比较可行的做法，但这种方法仅限于 100% 的实地，网点的定义只能依靠 CxF/X-4 数据。

6. 其他要求

代表印刷状态的 CMYK 色块应该打印在每一份打样中。标准中规定了应该包含的色块属性，如四色及叠印色实地、网点、灰平衡色块等，而这些色块都在 ISO 12642-2 中有定义。

通常情况下，并不需要自己制作控制条，但需要清楚所要匹配的目标标准，并选用对应的国际知名的印刷机构现成的控制条即可，譬如 IDEAlliance ISO 12647-7 Control Wedge 2013、Ugra/Fogra Media Wedge CMYK V3 等（这两个都是新版的 3 行控制条，不建议再使用旧版 2 行的控制条）。

一份数码样稿如果只有图文，没有任何附加信息是不规范的，往往会给使用者带来很大的困扰和麻烦。ISO 12647-7 标准详细规定了在打样上应该包含以下信息：

标记信息、执行标准版本 ISO 12647-7：××××；文件名字，数码打样系统名称，承印物材料种类，模拟的印刷条件，打样的时间和日期，测量条件，M0、M1 或 M2，着色剂种类，使用的色彩管理 ICC Profile，RIP 名称和版本，缩放比例，表面处理类型，最新校准的日期和时间，任何数据准备的详情，应用噪点信息等。

数码打样系统虽然比较稳定，但仍然会有颜色偏移。ISO 12647-7 标准规定，CMYK 实地及 50% 的网点，以及 RGB 需要被二次评估，最大色差 $\triangle E_{00}$ 不超过 2.0，应该选用同一个仪器，测量同一个位置，必要时需要重新校准打样系统。

除以上要求之外，还有耐磨性测试、光泽度要求、网点要求、无条痕、图像套印及分辨率等要求。

总之，一份标准的数码打样产品，需要尽量模拟实际的印刷色彩及视觉外观，还要有足够的打样信息以方便使用者。在耗材选用、仪器配置、测量模式、软件设定、色彩校正、标准选择上都要注意是否能满足客户要求，尤其在转换为 Fogra 51/52、GRACoL 2013 等新标准时更要注意打样系统的方方面面是否能达到新标准的要求。有了合格而准确的打样，才能更好地指导印刷生产，最大限度发挥出数码打样应有的作用。

第三节 测量仪器的使用方法

一、密度计

密度测量和密度计作为彩色复制品质量评定中最重要的专用仪器,无论是对工艺技术和原材料的评价,或者是在复制过程中用做检查、控制的手段;还是在生产过程中对彩色质量的鉴定和评价,都离不开密度计。

密度计的测量原理和印刷工人目视鉴定的原理很接近。图8-6为密度计测量工作原理示意。发自稳定光源1的光通过透镜2聚焦而射到印刷面上,有一部分光被吸收,吸收量取决于墨膜5的厚度和颜料的浓度,未被吸收的光由印刷纸张反射。透镜系统6收集与测量光线呈45°的反射光线并将之送到接收器(光敏二极管)8,光敏二极管将所接收到的光量转变为电量。电子系统9将此测量电流与基准值("标准白"的反射量)进行比较,根据此差值计算所测量墨膜的吸收特性。墨膜测量的结果显示于显示器10上。光路上的滤色镜4只允许印刷油墨相应波长的光线通过。

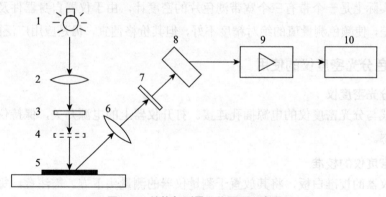

图 8-6 结构与测量工作原理示意图

用红、绿、蓝三种滤色片测量的密度,称为彩色密度或三滤色片密度,分别用 DR、DG、DB 表示。DR 反映了色料对入射光光谱中红光的吸收程度,同样 DG、DB 分别表示色料对入射光光谱中绿光、蓝光的吸收程度。因此,用彩色密度的三个独立参数 DR、DG、DB 可准确地表示某一色样的色彩属性。

二、分光光度计

分光光度计测量颜色表面对可见光谱各波长光的反射率。将可见光谱的光以一定步距(5 nm、10 nm、20 nm)照射颜色表面,然后逐点测量反射率。将各波长光的反射率值与各波长之间关系描点可获得被测颜色表面的分光光度曲线,也可将测得值转换成其他表色系统值,每一条分光光度曲线唯一地表达一种颜色。

分光光度计主要由光源(通常采用卤素钨灯或氙闪光灯,即标准光源A)、单色器(棱镜或光栅)、接收器(光电倍增管)和记录仪器(电位计)组成。其工作原理为,把光源的光分解成光谱,从所得的光谱中用狭缝挡板导出一单色光带,然后将单色光带投射到被测样品上去,单色光通过被测物体的透射或反射情况在记录表上被指示出来。对待测色来说,一般是从可见光谱 400～700 nm 的范围,每隔 10 nm 逐一测出光反射系数、吸收系数的透射系数,将各个波长的反射率或透射率用点连接起来便可绘出分光光度曲线。分光光度曲线可以表示有色物体完整色彩特征,一种彩色物体仅有一个分光光度曲线。

三、色度计

色度计是通过对被测颜色表面直接测量获得与颜色三刺激值 X、Y、Z 成比例的视觉响应,经过换算得出被测颜色的 X、Y、Z 值,也可将这些值转换成其他匀色空间的颜色参数。

色度计一般由照明光源、校正滤色器、探测器组成。其探测器是光电池、光电管或光电倍增管,它们的光谱灵敏度都经过滤色器的修正,以模拟 CIE 标准色度观察者的光谱三刺激值曲线。色度计用 3 个或 4 个探测器各自的相对光谱灵敏度曲线分别修正,由多个探测器的输出值得出待测色的三刺激值和色度坐标。色度计采用 45°入射光与 0°受光的方式进行测量,光源为卤素钨灯或氙闪光灯,即标准光源 A。

色度计实际上是一个带有三个宽带滤色片的密度计,由于仪器自身器件及原理方面存在一定的误差,使颜色测量值的绝对精度不好,但其价格便宜,仍是应用广泛的测色仪器。

四、彩色分光密度仪的使用

1. 连接分光密度仪

将电源线与分光密度仪的电源插孔连接。打开仪器上的电源开关,保持仪器的显示屏处于加亮状态。

2. 分光密度仪的校准

当使用仪器的校准白板,将其放置于测量仪器的测量头下方。需注意,每一台仪器对应一个校准白板,校准前需要确定所使用的白板的系列号与测量仪器的系列号是否一致。同时还要确定白板表面没有被污染,并保证测量仪器的测试区域与白板中心对应。

然后通过仪器上的菜单,选择测量仪器的校准(Calibration)功能,并按下测量头,等校准成功。

3. 密度测量(Density)

通过仪器面板上的选择(上/下)移动箭头,选择密度测量功能(按回车箭头)。测量密度时有相对密度与绝对密度的差别,相对密度指测量密度时将测量后的密度值减掉纸张密度后的密度值;绝对密度值指测量密度时直接测量样品所得到的密度值。两种不同类型的密度测量方式的选择可通过调整测量仪器上的可选项(Option)中的模式来实现。

同时，当需要比较样品与某一标准样的密度差值时，可以测量该标准样的密度值，并将其定义为参照样（Reference）。然后在测量时选择密度值与参照值相减（DEN—REF0X）的方式，就可以方便地得到样品与标准样间的密度差。此功能通过可选项（Option）中的参照项（Reference）来选择相关的所使用的参照标准的系列号，并需要改变密度测量功能的工作模式为密度—参照的方式（DEN—REF0X）。

4. 其他测量功能

分光密度仪除了可进行颜色样品密度的测量之外，还可测量印刷品的网点面积、印刷反差、叠印和色调误差与灰度等与印刷品质量测评相关的重要指标；同时，部分型号的分光光度仪还可测量颜色样品的色度值，如 CIEXYZ 与 CIELab 以及色差等。其他测量功能请参见相关品牌型号的设备的网站。

五、彩色分光光度计的使用

以 X-Rite Eyeone 仪器为例，介绍彩色分光计的使用方法。

1. 仪器连接与校准

将电源线与分光光度计的电源插孔连接，数据线与计算机的数据通信口（新款的分光光度计多使用 USB 接口与计算机连接）连接。

打开仪器上的电源开关，保持仪器的状态指示灯为加亮状态。

使用仪器的校准白板，将其放置于测量仪器的测量头下方或者测量特定的校准区域。与密度计同样，每一仪器对应一个校准白板，校准前需要确定所使用的白板的系列号与测量仪器的系列号是否一致；确定白板表面没有被污染，并保证测量仪器的测试区域与白板中心对应。通过应用程序的提示完成对仪器的校准。

2. 分光光度值的测量

分光光度计有多种工作方式，手动测量或自动扫描测量方式。

工作时通过使用测量软件，根据测量软件的指示首先选择仪器的型号，如果所选择仪器型号与连接的仪器型号一致，则状态将显示"OK"，确定测量的数据是颜色的分光光谱值或是色度值（测量值为分光光谱值即在仪器连接窗口中勾选 Spectral，反之则表示测量结果为色度值）。

然后，需要在测量软件中定义测量所需要的色表（Test Target），此色表可选择测量软件中已经定义的常用标准色表，也可通过用户自定义（Custom）的方式定义测量的颜色数量，以及每色块采样测量的次数。之后测量软件将提示将仪器对准校准板进行校准。

校准完成后，测量时使用自动扫描方式时，仪器将根据色表的定位结果计算出每次移动测量头的位置，并测量读出测量结果；使用手动测量的方式则需要用手动的方式将测量头对准色块，按下测量部分，完成测量。图 8-7 所示为测量软件 ProfileMaker 中的 MeasureTool 驱动分光光度测量的工作界面。

平版计算机直接制版技术

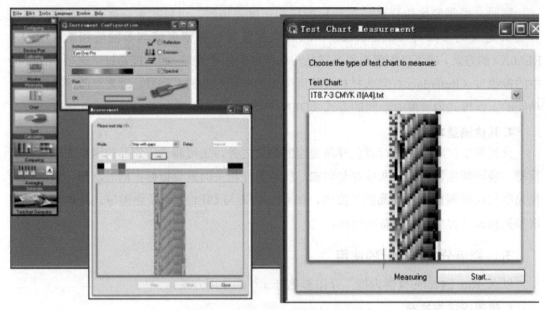

图 8-7　MeasureTool 测量工作界面

138

第九章 平版印刷标准与管理

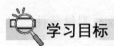

学习目标

1. 能进行印品的等级评定;
2. 能应用质量管理体系知识实现操作过程中的质量统计、分析与控制;能针对打样中可能出现的问题提出相应的预案;
3. 能依据 ISO-9001 标准制定打样工序的质量管理方案;
4. 能进行生产计划、调度、设备安全及人员的管理;
5. 能制定部门的环保作业措施;能制定、优化制版的工艺流程;
6. 能制定特殊工艺方案;
7. 能根据各工序生产情况制订生产计划;
8. 能分析产生质量问题的原因。

第一节 相关印刷质量标准

一、ISO 12647 标准

ISO 12647 印刷过程控制标准由国际标准化组织(International Standard Organization,ISO)内针对印刷技术的专门委员会 TC130 制定,其制定的国际标准主要可归纳为以下几大类:术语标准、印前数据交换格式标准、印刷过程控制标准、印刷原辅材料适性标准、人类工程学/安全标准。其中,印刷过程控制标准是印刷生产广泛应用的基础标准,在 ISO/TC130 中占有重要的地位,它主要规定了印刷生产过程中各关键质量参数的技术要求和检验方法。如 ISO 12647 的一系列标准，它是按照胶印、凹印、网印、柔印、数字印刷的工艺方法划分，针对不同印刷方式的质量控制参数，规定各自的技术要求和检验方法。

ISO 12647 标准（印刷技术、网目调分色、样张和印刷成品的加工过程控制）是在世界范围内多个国家印刷质量委员会的综合数据基础上发展而来的。它是一系列包装防伪标准，为各种印刷工艺（胶印，凹版印刷，柔性版印刷，等等）制作的技术属性和视觉特征提供最小参数集。该标准向制造商和印刷从业者提供了指导原则，有助于将设备设定到标准状态。来自这些印刷厂的测量数据可以用来创建 ICC 色彩描述文件并生成与印刷色彩相匹配的打样样张。

ISO 12647 是建立在包括油墨、纸张、测量和视觉观察条件标准之上的标准。这个标准包括很多部分。在 ISO 12647 标准的每一部分中，对不同的印刷工艺，定义了该工艺参数的最小值。

ISO 12647-1：网目调分色样张和印刷成品的加工过程控制：参数与测量条件，包装装潢

ISO 12647-2：平版胶印

ISO 12647-3：新闻纸的冷固型胶版印刷

ISO 12647-4：出版凹版印刷

ISO 12647-5：丝网印刷

ISO 12647-6：柔性版印刷

ISO 12647-7：直接来源于数字数据的样张制作，数码印刷和打样

简而言之，ISO 国际标准有助于印刷从业者、印前部门以及印刷品买家之间进行信息交流。作为国际上通用的标准，ISO 已经越来越广泛地被印刷品买家所接受和采用，也成为印刷企业进出口业务的通行证。

而在国内，采用国际标准的印刷企业也越来越多，如北京圣彩虹制版印刷技术有限公司、浙江影天印业有限公司等众多知名印刷企业，甚至一些规模还不是很大的企业，在印刷生产时均严格按照 ISO 12647-2、ISO 12647-7、ISO 2846、ISO 12646、ISO 10128 等一系列国际标准的要求去执行印刷流程的每一步操作，做到产品有序可循、有章可查，以确保产品的稳定性。

二、ISO 12647-2 标准

ISO 12647-2 是 ISO 12647 中的标准"印刷技术——网目调分色、打样和印刷的生产过程控制——第 2 部分：胶印机过程控制"，ISO 12647-2 的可信度是建立在实地密度块和 TVI（网点增大曲线）曲线的基础上。

1. 实地 CMYK 色度标准

随着印刷与检测技术的发展，印刷品质控制分析研究发现印刷密度指标不能准确地控制印刷四色油墨的品质，因此 ISO 12647-2 印刷过程控制标准于 2004 年发布的新标准中

将印刷四色实地色的测评指标修订为以 CIELab 色度值为标准的值，如表 9-1 所示。

表 9-1 ISO 印刷实地标准

纸张类型	1+2			3			4			5		
	L*	a*	b*	L*	a*	b*	L*	a*	b*	L*	a*	b*
黑色衬垫测量值（On Black Backing）												
黑色	16	0	0	20	0	0	31	1	1	31	1	2
青色	54	-36	-49	55	-36	-44	58	-25	-43	59	-27	-36
品红色	46	72	-5	46	70	-3	54	58	-2	52	57	2
黄色	88	-6	90	84	-5	88	86	-4	75	86	-3	77
红色（M+Y）	47	66	50	45	65	46	52	55	30	51	55	34
绿色（B+Y）	49	-66	33	48	-64	31	52	-46	16	49	-44	16
蓝色（C+M）	20	25	-48	21	22	-46	36	12	-32	33	12	-29
白色衬垫测量值（On White Backing）												
黑色	16	0	0	20	0	0	31	1	1	31	1	3
青色	55	-37	-50	58	-38	-44	60	-26	-44	60	-28	-36
品红色	48	74	-3	49	75	0	56	61	-1	54	60	4
黄色	91	-5	93	89	-4	94	89	-4	78	89	-3	81
红色（M+Y）	49	69	52	49	70	51	54	58	32	53	58	37
绿色（C+Y）	50	-68	33	51	-67	33	53	-47	17	50	-46	17
蓝色（C+M）	20	25	-49	22	23	-47	37	13	-33	34	12	-29

2. TVI 标准

在 ISO 12647-2 胶印印刷控制标准中对印刷网点增大进行了说明，表 9-2 表列出了 ISO 标准纸张与胶版印刷条件下的 50% 阶调时的网点增大值。图 9-1 为 ISO 根据特定印刷条件所获取的标准网点增大曲线。

表 9-2 ISO 标准纸张与胶版印刷条件下的 50% 阶调时的网点增大值

印刷条件	网点增大值（针对不同的加网线数）		
	52LPcm	60LPcm	70LPcm
四色彩色印刷各彩色版网点增大[①]			
阳图版[②]，纸张类型[③]1、2	17	20	22
阳图版，纸张类型 4	22	26	—
阴图版，纸张类型 1、2	22	26	29
阳图版，纸张类型 4	28	30	

(续表)

印刷条件	网点增大值（针对不同的加网线数）		
	52LPcm	60LPcm	70LPcm
轮转商业印刷等印刷条件下四色版网点增大[①]			
阳图版，纸张类型1、2	12	14（A）[④]	16
阳图版，纸张类型3	15	17（B）	19
阳图版，纸张类型4、5	18	20（C）	22（D）
阴图版，纸张类型1、2	18	20（C）	22（D）
阴图版，纸张类型3	20（C）	22（D）	24
阴图版，纸张类型4、5	22（D）	25（E）	28（F）

注：①黑版与其他彩色版的增大值相同或大3%。

②印版的种类相对直接制版技术应该是独立的，但在实际生产过程中往往使用不同的控制参数输出阳图版与阴图版。

③纸张的类型在 ISO 12647-2 胶印标准中有相关的规定。

④A、B、C、D、E、F 是图 9-1 所对应的标准曲线，曲线通过测量印刷复制品上的 CMYK 各色阶的色样的色度值计算后得到网点值，然后绘制曲线。

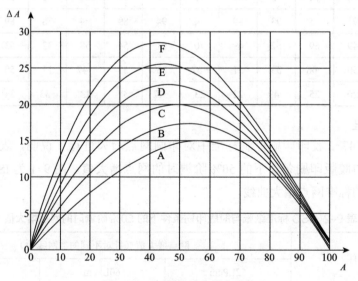

图 9-1　ISO 标准网点增大曲线

第二节　平版制版设备配置标准

一、CTP 网络环境

内存：2G 或以上；

硬盘：机械硬盘 7200r 或以上，硬盘大小 1T 或以上。软件运行盘（包括导入使用的 TIFF 文件）和系统盘请使用固态硬盘；

显卡：256M 或以上；

操作系统：Windows XP/Win7/Win8/Win10；

USB：2.0 Port；

CPU：Intel 双核 E5200/7400 或以上配置。

二、环境条件验收

制版机所在空间环境操作温度必须在：20～30℃，湿度：40%～60%。设备场地配置温度湿度计。

设备场地地面平坦度须在 ±4mm 以内。

设备准备专用地线，要求电压一定要稳定，地线直径 $4mm^2$（用镀锌金属直径不小于 10mm，深埋湿土 1.5m 以下），设备对地电阻≤0.5Ω，同时计算机机箱也要接地；准备一根一定长度的 $4mm^2$ 的接地线将设备与计算机外壳同时接地。

设备配置 UPS 不间断供电电源（>6kVA）。

设备外部连接配置 35A 的空开及标准的电源线。

设备周边不能有因振动会对制版机造成影响的大型设备。

工作间的除尘措施和达标要求应达到国家质量二级标准，即 API 值大于 50 且小于等于 100。

设备场地水质较差需安装净水装置，若水压不足，需安装水泵。

三、设备运输检查

设备到客户处后首先检查木箱外观有无破损、有无撞击、整体木箱的完整性是否良好。

拆除木箱后检查设备表面是否有油漆脱落、刮痕、罩壳凹陷、水痕、上盖移位等现象发生。

根据随机清单检查与设备配套软件、工具包、配件包、简易供版台等是否完整。

四、安装调试检查

制版机水平测试：制版机的三只升降地角升起保证地轮处于松弛状态，同时将水平尺置于光鼓表面，确认左右水平；再将水平尺放于墙板表面，确认前后水平。

冲版机水平测试：可根据显影槽的药液进行前后左右的水平调整。

确认所有固定扎带，固定装置拆除。

确认所有可活动部件的正常，如平衡块、压版辊等。

冲版机毛刷压力测试：剪一条 PS 版插入毛刷下方调整至压力适接触状态，左右压力一致。

冲版机胶辊压力测试：压力调整螺丝刚接触到胶辊轴套时，两边分别微量旋转，直至两边 90°左右。

冲版机功能测试：用温度计确认药水实际温度与显示温度一致；观察药水循环状态是否正常；添加制冷液，根据版材配套药水比例进行药水的添加，设置药水温度及冲洗时间，补充液浓度应高于显影槽的显影液配比。

五、制版机各项功能测试

（1）确认 UPS 不间断供电电源的正常连接。

（2）设备自检：设备执行相应动作时，工程师应严密观察设备相关机构有无异常动作及异常声响。

（3）确认所有参数均达到最佳值。

（4）激光测试：确认激光功率达到出厂标准。确认激光温度在正常范围（25℃±3℃）。

（5）曝光输出确认光鼓气压符合范围值。

（6）印版测试：确认网点范围值误差在 1% 以内；确认 1%～99% 网点还原；确认印版空白部分无底灰；确认图像叼口符合要求；确认图像居中，确认印版表面无任何划伤；确认四色套印误差在 0.01 mm 以内。

六、印版质量问题产生的原因及解决方案

1. 曝光能力不足或过大，使网点增大或损失

解决方案：采用能量测试，选择最佳能量。

2. 冲洗不足或过冲，印版产生底灰或者网点丢失

解决方案：调节冲洗时间，做好显影液补充，定期更换药水。

3. 印版有伤痕

解决方案：可能印版本身有伤痕，更换印版；可能冲版机胶辊造成，通常是胶辊表面结晶划伤，或者胶辊老化变形；清洗胶辊或者更换变形的胶辊。

4. 叼口位置不对，或者图像不居中

解决方案：校准叼口参数和图像参数。

5. 图像错位

解决方案：通常外部干扰造成，确认接地状况良好，电源线和数据线分开捆扎。

6. 网点发虚非能量和药水问题

解决方案：镜头或者光电池表面污染，进行清洁处理，或者焦距参数偏离，校准焦距参数。

7. 四色套位不准

解决方案：调整侧规重复定位精度，使其定位精度达到 0.01mm。

8. 图像在实地部分出现白线或者在空白部分出现黑线（指药膜的颜色）

解决方案：可能某路激光不受控制，更换激光控制板。

9. 印版曝光一部分然后停止曝光

解决方案：可能数据传输有问题，更换 USB 接口，或者更换 USB 线。

第三节 工艺流程控制

G7 印刷认证由美国的平版胶印商业印刷规范组织（GRACoL），结合 CTP 设备的多年实践，探索总结而成，其目的就是要在 CTP 的帮助下，实现商业胶印品质的一致化效果。

一、准备工作

预计的时间长度和工作过程共需要两次印刷操作，分别为校正基础印刷和特性化印刷，各需一到两个小时，中间需要半个小时到一个小时的印版校正，共约需半个工作日。所有的工作都应安排在同一天，并由相同的操作人员对同样的设备材料来完成。

1. 设备

印刷机调试到最佳工作状态，包括耗材，并检查其相关的物化参数是否符合要求。按生产厂家的要求，调节 CTP 的焦距、曝光及化学药水，并使用未经线性化校正的自然曲线出版。

2. 纸张

使用 ISO 1# 纸，尽量不带荧光。纸张需要 6000～10000 张不等，由操作效率决定。

3. 油墨

使用 ISO 2846-1 油墨。

4. 标准样张

可以从 www.printtools.org 网站上购买预置好的《GRACoL7 印刷机校正范样》，也可以自己做，如图 9-2 所示。

标准样张应该包括：

（1）两份 P2P23× 标准（或较新的版本），且互呈 180°；

（2）GrayFinder20 标准（或较新的版本）；

（3）两张 IT8.7/4 特性标准样，相互呈 180°，且排成一排；

（4）一条横布全纸张长的半英寸（1 厘米）（50C、40M、40Y）的信号条；

（5）一条横布全纸张长的半英寸（1厘米）50K的信号条；

（6）一条合适的印刷机控制条，应包括G7的一些重要参数，如HR、SC、HC等；

（7）一些典型的CMYK图像。

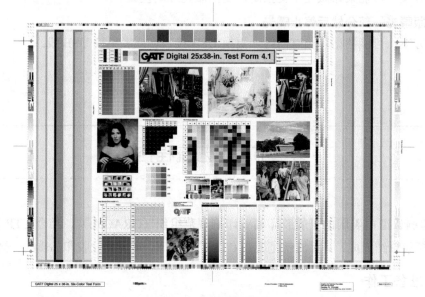

图9-2 标准样张示意

5. 其他

其他设备包括有由GRACoL免费提供的NPDC图纸（见图9-3）；测量印版的印版网点测量计；分光光度计；D50标准观察光源；做图用的曲线工具等。也可以购买GRACoL的软件IDEAlink来帮助快捷完成测试工作。

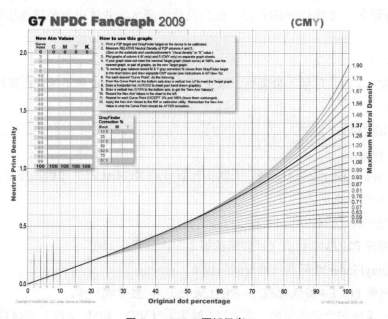

图9-3 NPDC图纸示意

二、校正基础印刷

1. 印刷条件

印刷机及其耗材都应得到正确调节，包括油墨的黏性、橡皮布、包衬、压力、润版液、环境温度、湿度等。印刷的色序建议为 K-C-M-Y。最好不使用机器的干燥系统。

2. 实地密度（SID）

按标准实地油墨的色度值（L*a*b*）或密度值印刷（如表 9-1 所列的数据）。

3. 网点增大曲线

测量每一 CMYK 色版的 TVI。CMY 的每条网点增大曲线之间的差值应在 ±3% 之内，黑版略高 3%～6%。

4. 灰平衡

将分光光度计设定在 D50/2°，测量印张上的几个 HR（50C、40M、40Y）块的灰平衡值。显示针对 L*a*b* 的误差组合情况，调节 CMY 的油墨实地密度。

如果灰平衡不能接近目标 a*、b* 值，或者不能通过少量实地密度的调节来改正，那么，请检查一个或多个 TVI 是否过大，油墨的色相是否不对，叠印是否不好（可能是由于油墨黏性 TACK，或乳化不好），或者是油墨色序不对。

5. 调节印刷均匀性

调节印刷均匀性可能是印刷机校正中最难的部分。调节印刷机墨键，尽量减小印张上实地密度的偏差，最好每种油墨在印刷面上的偏差不要超过 ±0.05，才能使灰平衡的偏差尽可能小，最好在印刷区域上或是滚筒处不超过 ±1.0a* 或 ±2.0b*。

6. 印刷速度

用每分钟 1000 张的生产速度运行机器，再次检查实地密度、灰平衡和均匀性。如果油墨的实地密度、灰平衡或均匀性的变化超过了数值，调节印刷机，确保得到希望的印刷要求，然后按需要再次提速，以正常的生产速度印刷，保证印刷品质量的均匀稳定。

三、CTP 的校正

1. 三色 CMY 曲线的校正

进行完第一次印刷后，检查 P2P 的第四列的色度值，中性灰有可能会做得很好，也可能不好。

（1）灰平衡达到后

①测量 P2P：选出符合要求的印刷品，干燥后测量 P2P 的第四列的数值的相对密度值。注意，应从不同区域至少测两个读数，取其平均值。

②绘出实际的 NPDC 曲线：在 GRACoL 的官方网站下载免费的图纸，做出印刷输出实际的 NPDC 曲线图。

③确定标准曲线：在 NPDC 扇形图中，找到最接近实际生产的实地密度值的目标图。若没有，可从上下两条接近的曲线中，自行用曲线板分析画出。

④确定校正点：检查实际曲线，看看在哪儿弯得最明显，然后确定需要校正的曲线点。由于人眼对亮调较敏感，因此，可以在亮调处多设几个点。

⑤校正 NPDC 曲线：在每一个校正点从（60，44）处往上画一条竖线，与目标线相交并从交点处再画一条横线，（向左或向右）与标准线相交从交点处向下画一条竖线，交于坐标轴，将获得一个新的目标值。在图纸上记录该值，并在每一曲线点处重复上述步骤，需注意 0% 和 100% 处不要动。

（2）未达到灰平衡

① GrayFinder：若印刷时不能实现想要的灰平衡，就可以用 GrayFinder 来完成校正。用一台分光光度计，测量标定青色为 50% 处（实际是 49.8%）色块的中间，以及相邻的色块，寻找一个最接近目标的中性灰值（即 0a*，-1b*）。如果中间色块最接近目标灰，那么，该设备已经灰平衡了（在 50%C 处），不需要做任何校正。如果最靠近目标的 a*b* 值不是中间的色块，注意 M 和 Y 旁边所列的百分数值。例如，如果最佳测量在 +2 和 +3 的 M 之间，-3Y 上得到，那么所要的较好的灰平衡为 +2.5M 和 -3Y。重复此步骤，可以为 175%、62.5%、37.5%、25% 和 12.5% 等色块，找到实际的灰平衡数值。

②确定单色 C、M、Y 的 NPDC 曲线：在 CMY 图上，先画出 P2P 的第四列值的曲线，即为 C 版曲线，然后通过在 GrayFinder 上所找到的百分数，画出单色 M 和 Y 版的曲线，在原曲线（C 版）的左边或右边。

③确定标准的 NPDC 曲线、校正点并画出新的 NPDC 曲线。

2. 确定印刷实地密度值的标准曲线

在 NPDC 扇形图中，确定最接近实际印刷实地密度值的标准曲线。然后，找到需要校正的点。对每一个校正点进行校正。从下至上画一条竖线，交于目标线在交点处，画一条横线，与 C、M 和 Y 线相交在交点处，从上至下画若干条竖线，与坐标横轴相交，得到 C、M、Y 的三个新的目标值。在图纸的新值栏上记录 CMY 的目标值重复每一个曲线点，0% 和 100% 不变。

3. 单色黑版的校正

在黑版专用图纸上，将 P2P 第五列的数值绘上。做法可参考三色 CMY 的 NPDC 校正。

4. 为 RIP 赋值

将上述 CMY 和 K 的 NPDC 校正结果，为 RIP 或校正设备赋新的目标值。有些 RIP 设备需要输入"测量后"的值，而不是"所需要"的值，还有些 RIP 需要输入校正的差值。新的目标值就是经过校正后每个曲线点都应该得到的值。

四、印刷

1. 制版

用新的 RIP 曲线，制作标准样张的新印版，并且将 P2P 上的印版值与所记录的未校正过的印版曲线进行对比，确保所要求的变化已经获得。例如，如果 50% 的曲线点有一个新的目标值为 55%，则检查新版的 50% 色块处是否比未校正过的印版大约增大 5%。由于印版表面测量困难，因此，只要这些值大概正确即可。

2. 印刷

使用新的印版或 RIP 曲线，并且用相同的印刷条件，印刷特性标准样，最好是整个测量标准样。照着与校正印刷最后所记录相同的 L*a*b* 值（或密度值）来印刷。注意墨色的均匀性和灰平衡。调机时，测量 HR、SC 和 HC 值，确定印刷机满足 NPDC 曲线。如果可能的话，也测量 P2P 标准样，或者在一张空的 G7 图纸上动手绘出第 4 和第 5 列。这些曲线此时应该可以与目标曲线几乎完全一致。如果不是，调节实地密度，或者再多印几张让印刷机预热。检查其他参数，如灰平衡、均匀性等，其数据都在控制内，然后高速开动机器到正常印刷速度，检查测量值，看看是否到最后都很好。从现场选取至少两张或更多张，自然干燥。如果可能，再以相同的条件，进行两次或更多次的印刷机操作，从每一次印刷中选取最佳的印张，为后续工作平均化准备。

3. 建立 ICC

用分光光度计，测量所选取的每一印张的特性数据，然后从平均数据中建立印刷机的 ICC 文件。如果可能，存储原来测量的光谱数据，而不是 CIEL*a*b*（D50）的数据。如此得到改良后的 ICC 文件，可以减少由于非标准光源，或是两种光源的变化而导致的同色异谱的问题。

第四节　生产与环境管理

一、相关知识

1. 相关环境管理体系标准

ISO 14000 系列环境管理标准是国际标准化组织（ISO）在成功制定 ISO 9000 族标准的基础上设立的管理系列国际标准。目前 ISO 14000 标准的最新标准是 2015 年版，我国等同采用该标准并颁布了 GB/T 24001-2016《环境管理体系要求及使用指南》。

（1）ISO 14000 系列标准的组成

ISO 14000 系列标准是国际标准化组织 ISO/TC 207 负责起草的一份国际标准。ISO 14000 是一个系列的环境管理标准，它包括了环境管理体系、环境审核、环境标志、生命周期分析等国际环境管理领域内的许多焦点问题，旨在指导各类组织（企业、公司）取得和表现正确的环境行为。

ISO 14001 标准是 ISO 14000 系列标准的龙头标准，它的总目的是支持环境保护和污染预防，促进环境保护与社会经济的协调发展。为此，ISO 14001 标准突出强调了污染预防和持续改进的要求，同时要求在环境管理的各个环节中控制环境因素、减少环境影响，将污染预防的思想和方法贯穿环境管理体系的建立、运行和改进之中。目前，现代企业正在推行实施 ISO 14001 环境管理体系标准。

（2）ISO 14000 系列标准的作用与意义

①有助于提高组织的环境意识和管理水平

ISO 14000 系列标准是关于环境管理方面的一个体系标准，它是融合世界许多发达国家在环境管理方面的经验于一身而形成的一套完整的、操作性强的体系标准。企业在环境管理体系实施中，首先对自己的环境现状进行评价，确定重大的环境因素，对企业的产品、活动、服务等各方面、各层次的问题进行策划，并且通过文件化的体系进行培训、运行控制、监控和改进，实行全过程控制和有效的管理。同时，通过建立环境管理体系，使企业对环境保护和环境的内在价值有进一步的了解，增强企业在生产活动和服务中对环境保护的责任感，对企业本身和与相关方的各项活动中所存在的和潜在的环境因素有充分的认识。该标准作为一个有效的手段和方法，在企业原有管理机制的基础上建立起一个系统的管理机制，新的管理机制不但提高环境管理水平，而且还会促进企业整体管理水平。

②有助于推行清洁生产，实现污染预防

ISO 14000 环境管理体系高度强调污染预防，明确规定企业的环境方针中必须对污染预防作出承诺，推动了清洁生产技术的应用，在环境因素的识别与评价中全面地识别企业的活动、产品和服务中的环境因素。对环境的不同状态、时态可能产生的环境影响，以及对向大气、水体排放的污染物、噪声的影响以及固体废物的处理等逐项进行调查和分析，针对现存的问题从管理上或技术上加以解决，使之纳入体系的管理，通过控制程序或作业指导书对这些污染源进行管理，从而体现了从源头治理污染，实现污染预防的原则。

③有助于企业节能减排，降低成本

ISO 14001 标准要求对企业生产全过程进行有效控制，体现清洁生产的思想，从最初的设计到最终的产品及服务都考虑了减少污染物的产生、排放和对环境的影响，能源、资源和原材料的节约，废物的回收利用等环境因素，并通过设定目标、指标、管理方案以及运行控制对重要的环境因素进行控制，达到有效地减少污染、节约资源和能源，有效地利用原材料和回收利用废旧物资，减少各项环境费用（投资、运行费、赔罚款、排污费），从而明显地降低成本，不但获得环境效益，而且可以获得显著的经济效益。

④减少污染物排放，降低环境事故风险

由于 ISO 14000 标准强调污染预防和全过程控制。因此，通过体系的实施可以从各个环节减少污染物的排放。许多企业通过体系的运行，有的污染物被替代避免了污染物的排放；有的通过改进产品设计、工艺流程以及加强管理，减少了污染物的排放；有的通过治理，使得污染物达标排放。实际上 ISO 14000 标准的作用不仅是减少污染物的排放，从某种意义来说，更重要的是减少了责任事故的发生。因此，通过体系的建立和实施，各个组织针对自身的潜在事故和紧急情况进行了充分的准备和妥善的管理，可以大大降低责任事故的发生。

⑤保证符合法律法规要求，避免环境刑事责任

现在，世界各地各种新的法律法规不断出台，而且日趋严格。一个组织只有及时地获得这些要求，并通过体系的运行来保证符合其要求。同时由于进行了妥善有效控制和管理，可以避免较大的事故发生，从而避免承担环境刑事责任。

⑥满足顾客要求，提高市场份额

虽然目前 ISO 14000 标准认证尚未成为市场准入条件之一，但许多企业和组织已经对供货商或合作伙伴提出此种要求，一些国际知名公司鼓励合作公司按照 ISO 14001 的要求，比照自己的环境管理体系，力争取得对这一国际标准的注册，暗示将给予正式实施 ISO 14001 的供应商以优先权。

⑦取得绿色通行证，走向国际贸易市场

从长远来看，ISO 14000 系列标准对国际贸易的影响是不可低估的。目前，国际市场上兑现的"绿色壁垒"多数是由企业向供货商提出的对产品或生产过程的环境保护要求，ISO 14000 系列标准将会成为国际贸易中的基本条件之一。

实施 ISO 14000 系列标准将是发展中国家打破贸易壁垒，增强竞争力的一个契机。

ISO 14000 系列标准为组织，特别是生产型企业提供了一个有效的环境管理工具，实施标准的企业普遍反映在提高管理水平，节能降耗，降低成本方面取得不小的成绩，提高了企业产品在国际市场上的竞争力。

（3）ISO 14001: 2015 环境管理体系标准

见相关标准。

2. 制版设备利用率和成本控制方法

设备利用率是表明设备在数量、时间和生产能力等方面利用状况的指标。提高设备利用率是提高印刷企业经济效益的重要目标，而提高设备利用率的措施则是实现这一目标的重要内容。

一般而言，制版设备利用率低下的因素有：制版故障造成的停机；制版设备调整的非生产时间（即制版准备时间）；设备空转和短暂停机；生产有缺陷的产品和降低设备的产量（开机造成的损失）。为全面提高制版设备生产利用率，可从实现制版设备效率的最大化，制订彻底有效的制版设备预防性维护计划，建立持续改进制版设备利用率的实施小组，在企业内实施全面生产质量管理和激励政策等四个方面着手。

制版设备成本不但包括一次性购置设备的付出，还包括使用过程中的维护、保养、维修等。制版设备成本控制就是控制制版设备的购置、使用、维护、维修和能力的充分发挥。可从以下几个方面加强管理。

(1) 制版设备引进的决策要正确

设备引进决策正确与否是印刷企业控制成本的第一关键条件和前提。印刷企业在购进制版设备时，要综合分析拟购设备的性能、价位和配件、维修、服务成本等，设备要适应所在印刷市场的活源与供求关系，以提高设备的开工率，减少停机成本。还要了解该设备的寿命周期，在设备的优势年龄段购入。

(2) 要做好制版设备的日常运行和维护

企业设备管理部门或生产部门要制定制版设备的日常管理、维护制度，并对车间生产实施过程进行检查监督；日常的维护要有书面制度，如定时定位查看、清洁等；制定周、月、年保养的具体范围和内容；还要灵活掌握保养的时间，忙时可以适当延期，闲时就要仔细检修，一定要从设备方面做好充分准备，防止关键时刻掉链子。

(3) 易损件的购买与准备要形成三级标准目录

配件、易损件应分别按不同时限提前购进，并按存货的最低数量随时补充，以保证生产的不间断进行；非易损件紧急损坏时的处置要有预案，要有最短的供货途径和时间；最好有不止一个以上的供货渠道，做到双保险。

(4) 应把设备的设计能力发挥到极限

为适应市场激烈的竞争，印刷企业普遍缩短了印刷设备的折旧年限，因此，要求我们在有限的时间内，要充分发挥出制版设备的生产能力。要在设备调试、磨合好以后，依据产品情况，在许可的前提下，尽量提高制版设备使用效率。

3. 节能减排的管理知识

广义而言，节能减排是指节约物质资源和能量资源，减少废弃物和环境有害物（包括三废和噪声等）排放；狭义而言，节能减排是指节约能源和减少环境有害物排放。

我国早在1997年制定、1998年正式施行的《中华人民共和国节约能源法》，将节能赋予法律地位。内容涉及节能管理、能源的合理利用、促进节能技术进步、法律责任等。该法明确了我国发展节能事业的方针和重要原则，确立了合理用能评价、节能产品标志、节能标准与能耗限额、淘汰落后高耗能产品、节能监督和检查等一系列法律制度。

印刷企业作为服务加工型企业，并非工业企业中的耗能大户，因此印刷企业的节能关键是在生产的各个能源使用环节上要减少损失和浪费，提高其有效利用程度。印刷企业主要节能措施有：

(1) 热能节约

热能主要使用形式是为印刷工艺过程某环节加热、对原料和产品的热处理、企业建筑冬季采暖等，节能途径的关键是提高热交换过程的效率、尽可能使用低晶位的热能，特别是余热。如印刷企业的印前制版显影、胶印油墨干燥、凹印色间干燥、无线胶订上胶等环节，都是可以节约热能、利用余热的环节。

(2) 电能节约

电能作为全球应用最广的二次能源，已得到普遍应用。但在传输和使用中不可避免地会有损耗。提高输电效率、提高用电设备的利用率，将对节约能源起到重要作用。印刷企业作为终端用电户，节电措施应为淘汰低效电机或高耗电设备，改造原有电机系统调节方式，推广变频调速、独立驱动等先进用电技术，正确选择电加热方式，降低电热损失。

（3）水能节约

水资源短缺是我国尤其是北方地区经济社会发展的严重制约因素，中国已开始进入用水紧张时期。印刷企业在生产中的胶片显影、印版显影、印刷机循环冷却、印刷车间冷却等都较大量用水，提高用水设备的能源利用效率，采用新工艺降低产品生产的有效用水，从而能够直接节约水能。

（4）降低能耗

加强企业科学的组织管理，通过各种途径减少原材料消耗，如纸张、油墨、润湿液、版材、胶片、胶辊、橡皮布、洗车水、用胶、薄膜等，在保证印刷品质量的前提下，既要减少印刷直接能耗，也要减少印刷间接能耗。

（5）提高印刷设备利用率

先进印刷设备的投资巨大，每年的设备维护、维修费用也不小，如何充分利用好印刷设备，发挥好印刷设备的全部功能，使其达到最大限度地使用，减少印刷设备的非工作时间，特别是还要付出成本的维修，就是充分利用了印刷设备。

二、操作步骤

（1）了解产品加工工艺流程和产品质量的要求。

（2）对全面质量管理诸多因素进行分析与控制。

（3）在实施过程中，找出需要调整或重点加以控制的因素，稳定产品质量。

（4）实施全面质量控制和综合管理。

三、注意事项

（1）各从业人员应积极学习生态环境部印发的《国家环境保护标准"十三五"发展规划》。

（2）各单位应开展多种形式的宣传教育活动，普及绿色印刷知识，增强全行业从业人员的绿色印刷意识。

参考文献

[1] 穆健. 实用电脑印前技术 [M]. 北京：人民邮电出版社，2008.

[2] [德]Helmut Kipphan. 印刷媒体技术手册 [M]. 谢普南，王强，译. 广东：世界图书出版公司，2004.

[3] 田全慧，张建青，莫春锦. 印刷色彩控制技术 [M]. 北京：印刷工业出版社，2014.

[4] 赵广，姚磊磊. ISO 12647-7 数码打样新标准话你知 [EB/OL].(2017-10)[2023-10]. http://www.keyin.cn/people/mingjiazhuanlan/201710/30-1107622.shtml.

[5] 沈伟志. 数码打样色彩管理认证 [J]. 数码印刷，2010.

"The Fourteenth Five-Year Plan" textbook for printing majors of higher vocational college

The bilingual training material of lithographic platemaking

Computer to Plate of Lithographic

Chief Editor | Chen Bin Xiang Jianlong

文化发展出版社
Cultural Development Press
· 北京 ·

图书在版编目（CIP）数据

平版计算机直接制版技术：汉、英 / 陈斌，项建龙主编. — 北京：文化发展出版社，2024.5
ISBN 978-7-5142-4336-9

Ⅰ.①平… Ⅱ.①陈… ②项… Ⅲ.①计算机辅助制版－汉、英 Ⅳ.① TP391.72

中国国家版本馆 CIP 数据核字 (2024) 第 080993 号

平版计算机直接制版技术（中英文）

主　　编：	陈　斌　项建龙
出 版 人：	宋　娜
责任编辑：	李　毅　杨　琪　雷大艳　　责任校对：岳智勇
责任印制：	邓辉明　　　　　　　　　　封面设计：韦思卓
出版发行：	文化发展出版社（北京市翠微路 2 号　邮编：100036）
发行电话：	010-88275993　010-88275710
网　　址：	www.wenhuafazhan.com
经　　销：	全国新华书店
印　　刷：	北京九天鸿程印刷有限责任公司
开　　本：	787mm×1092mm　1/16
字　　数：	450 千字
印　　张：	21.75
版　　次：	2024 年 6 月第 1 版
印　　次：	2024 年 6 月第 1 次印刷
定　　价：	68.00 元

ISBN：978-7-5142-4336-9

◆ 如有印装质量问题，请与我社印制部联系。电话：010-88275720

Foreword

The Chinese civilization has a glorious history over five thousand years with countless brilliant achievements. As one of the shining stars in the excellent traditional Chinese culture, printing has played a critical role in promoting the inheritance and development of human civilization. With the changes of times and the development of technology, China's printing industry has stepped into a new era of high-quality development, and is undergoing transformation and upgrading towards "greenization, digitization, intelligence and integration". In recent years, there have been numerous innovations in printing technology and craftsmanship in China, providing strong impetus for the sustainable development of Chinese printing industry.

The upgrade of printing technology is based on the innovation of platemaking technology. With the rise and rapid development of Artificial Intelligence Technology, the platemaking technology is rapidly developing towards intelligence, and will become one of the key areas for the intelligent development of printing technology in China in the next decade. The Chinese platemaking enterprises, aiming for technological changes and industrial optimization and upgrading, driven by technological innovation, seeking survival with product quality, constantly make progress and create intelligent platemaking system directly controlled by computer. With high-quality innovation, stable product characteristics and thoughtful after-sales services, they provide new impetus for the development of the global printing industry. At present, the technology and service capability of China's platemaking equipment have reached the international advanced level, taking a place in the international market.

For the purpose of adapting to the trend of digital development in printing technology, meeting the demand of the global printing industry for skilled talents with high-quality and high-level platemaking technology, better serving the training of skilled printing talents, and promoting the development and popularization of intelligent platemaking technology, Shanghai Publishing and Printing College and Hangzhou Cron Machinery & Electronics Co., Ltd. have jointly developed the textbook *Computer to Plate of Lithographic* by leveraging their respective advantages. The content of this textbook includes the types of lithographic plates, platemaking technology, Computer to Plate (CTP) platemaking, post-processing for plate development, etc., covering various technical process control links of platemaking of lithographic. This textbook development focuses on the cultivation of professional skills in the field of lithography and platemaking, with reference to the Chinese national vocational technical standards of the *Prepress Processors and Producers* and *the Printing Equipment Maintenance Workers* (CTP). It closely follows the key operations of lithography and platemaking jobs, focuses on shaping basic professional technical operating and comprehensive technical application abilities, and is suitable for those professional workers engaged in lithographic printing and platemaking work with certain basic professional printing knowledge.

This textbook has been jointly completed by Chinese printing vocational education institutions and Chinese innovative and high-quality printing enterprises. The textbook has integrated the leading technology with rich contents. It will continuously lead the industry's development with technology, knowledge and professional spirit, and help to promote the digital and intelligent upgrade of the printing industry. Especially, this bilingual book is more conducive to promoting personnel communication, technical exchange and digital connectivity in the global printing industry. Here, we sincerely hope that printing companies and professionals from all over the world will take the chance to know this textbook, actively participate in exchanges in the field of printing talent education, establish long-term and close cooperation mechanisms, and jointly promote the development of the global printing industry to create a better future.

Liu Binjie

May, 2024

Preface

Platemaking of lithographic refers to the process of making an original manuscript into a lithographic plate. The principle of lithographic is based on the insolubility of oil and water: the blank area with favorable hydrophilicity can repel ink after absorbing water while the lipophilic printing area can repel water and absorb ink; Wet the lithographic plate with water first to allow the blank area to absorb moisture, and then apply ink. The blank area absorbs water while the printing area absorbs ink, whick form graphic and textual information by the transfer of ink. The commonly used lithographic plates include PS plates (Pre-Sensitized plates), deep-etch plate, protein plates (dry offset plates), multi-layer metal plates, etc. The surface of each type of lithographic plate is composed of hydrophobic graphic and textual areas and hydrophilic blank areas. The process of platemaking of lithographic has evolved from the initial technology based on stone platemaking to a very mature technology after continuous development.

Computer to Plate of Lithographic is a digital lithographic imaging technology that emerged in the 1990s. At the Germany Drupa Printing Technology and Equipment Exhibition (Drupa) in 1995, there were 42 CTP systems exhibited for the first time. The CTP technology adopts a digital workflow, which directly converts text and images into numbers, controls them directly with a computer, and generates a lithographic plate directly printable on machine with laser scanning imaging and development processing. The CTP technology has eliminated the process of manual typesetting and plate buming. Applying computer and digital technology, CTP has a number of advantages including fast speed, high efficiency, fewer processes, simplified procedures, less variables, full-process digital processing, improved stability and standardized management.

This textbook is an important part of the Chinese national professional technical standard for the *Prepress Processors and Producers*. It consists of nine chapters, and focuses on the ability of platemaking of lithographic jobs. It comprehensively explains the CTP process and technology from the basic knowledge of platemaking of lithographic, the process of CTP, the key points of platemaking equipment operation, and the management of sampling and standardization. This textbook emphasizes the concept that knowledge serves operational skills, has strong practicality, and is suitable for professionals engaged in platemaking of lithographic with certain basic knowledge in printing. It can also serve as technical guidance for platemaking of lithographic as well as a reference book for professional assessment and learning in platemaking of lithographic. We hope that after reading this book, readers will have a thorough and comprehensive understanding of the CTP technology, and improve their overall understanding of platemaking of lithographic technology by combining it with practical applications in the production.

This is a bilingual textbook to be published coincided with the eleventh anniversary of the the Belt and Road Initiative formulated by Xi Jinping, the President of PRC. For the last eleven years, "The jointly constructed Belt and Road Initiative has become a popular platform of international public goods and cooperation," Xi Jinping underlined in the report of the 20th National Congress of the CPC, and emphasized that we needed continue to "promote the quality development of 'the Belt and Road.'" The Belt and Road Initiative adheres to the principle of extensive consultation and cooperation to achieve common growth, and will guide China's printing industry to play its due role in actively promoting the international cooperation.

We would like to thank Wang Dan, Tian Quanhui and other teachers from Shanghai Publishing and Printing College who has given great help and support in the compilation of this textbook, and Chen Wei, Fan Xiejun, Li Chundi, Victor Wong, Yao Wen, Dong Hai, Zhang Lixian and other personnel from Hangzhou CRON Machinery & Electronics CO., LTD. who actively participated in the compilation and publication of this textbook.

It is beneficial to read a book. We hope this textbook will provide some practical help to readers. At the same time, readers are welcome to put forward valuable comments and suggestions on the shortcomings of the book, so as to supplement and correct this textbook during revision!

Editor

Spring, 2024

Contents

Chapter 1 Introduction

Section 1 Classification and Process of Lithographic Platemaking / 1
Section 2 Principle and Process of Traditional PS Plate Platemaking / 2
Section 3 Principle and Process of Lithographic CTP / 4

Chapter 2 Basic Knowledge of Lithographic Platemaking

Section 1 Basic Working Principles of CTP Machine / 16
Section 2 Output Resolution and Its Setting / 18
Section 3 Role of Screening and Its Setting / 19
Section 4 Lithographic Platemaking Output File Format / 21
Section 5 Type Technology / 26
Section 6 Basic Concepts and Methods of Imposition / 30
Section 7 CTP Screening Technology / 32
Section 8 Basic Knowledge of Printing Compensation and Inverse Compensation Curves / 36

Chapter 3 Printing plate output preparations

Section 1　Environment Requirements for Production Operation ／ 40
Section 2　Installation of Output Process Software and Device Driver Software ／ 42
Section 3　Linearization of Printing Press Output ／ 45
Section 4　Operation of Output Process Software ／ 49
Section 5　Platemaking Machine and External Equipment ／ 69
Section 6　Processor ／ 75

Chapter 4　Output of Printing Plate

Section 1　CTP Printing Plate Output ／ 78
Section 2　Set the Resolution and Screening Parameters of the Press ／ 88
Section 3　Solutions to Faults Arising in the Process of Plate Output ／ 93

Chapter 5　Development of Printing Plate

Section 1　Development of CTP Printing Plate ／ 99
Section 2　Fault Resolution to Development of CTP Printing Plate ／ 102

Chapter 6　Quality Inspection of Printing Plates

Section 1　Common Quality Issues with CTP ／ 109
Section 2　Printing Plate Instruments ／ 110
Section 3　Printing Plate Output Inspection ／ 112
Section 4　Quality Control of CTP Printing Plate Imaging ／ 114
Section 5　Principle and Control of Dot Enlargement ／ 118

Chapter 7　Test and Anomaly Handling of Platemaking Equipment

Section 1　Printing Plate Output and Test of CTP ／ 123
Section 2　Analysis and Handling of Anomalies ／ 127
Section 3　Methods for Judging the Quality of Printing Plate Output ／ 133
Section 4　Prepress Processing and the Relationship between Platemaking and Printing Quality ／ 139

Chapter 8　Checking Digital Proofs

Section 1　Proof Quality Inspection and Control Methods ／ 141
Section 2　Digital Proofing Measurement Control Strip ／ 144
Section 3　Method of Application of Measurement Instruments ／ 151

Chapter 9　Lithographic Printing Standards and Management

Section 1　Standards of Related Printing Quality ／ 155
Section 2　Configuration Standards of Lithographic Platemaking Equipment ／ 159
Section 3　Process Control ／ 161
Section 4　Production and Environment Management ／ 167

References

Chapter 1 Introduction

Objectives:

1. Master the basic concepts and methods of lithographic platemaking;
2. Understand the basic process flow and characteristics of traditional PS plate platemaking;
3. Understand the process flow and characteristics of CTP;
4. Master the basic process of CTP platemaking.

Section 1 Classification and Process of Lithographic Platemaking

Lithographic platemaking is the process of making the original into a printing plate of lithographic printing. The printing part and the blank part of the lithographic printing are almost on the same plane. The blank part has good hydrophilic performance, which can exclude the ink after absorbing water, while the printing part is lipophilic, which can exclude the water and absorb the ink. When printing, the printing plate is first wetted with water to make the blank part absorb moisture; then the printing ink is applied. The blank part has adsorbed water and can no longer absorb the ink, while the printing part adsorbs the ink. The printing part on the printing plate can perform print with ink. Typically, there are Pre-Sensitized (PS) plates, deep-etch plates, protein plates (dry offset plates), multi-layer metal plates and so on. The surface of each printing plate is composed of the lipophilic graphic part and the hydrophilic blank part.

There are many different methods of lithographic platemaking depending on the materials and methods used.

I. Classification

1. Classified by Printing Plate Material

The plates of lithographic platemaking are of types such as slate, zinc, aluminum, paper base, etc.

The slate is the earlist lithographic printing material. The slate is cumbersome so it is only suitable for direct printing, and that's why it is no longer used now. Zinc and aluminum plates are commonly used today. The paper base has been developed in recent years and is often used in electrostatic printing, small-number-of-prints printing and light printing.

2. Classified by Printing Method

Lithographic platemaking is classified into the hand-drawn plate, transfer plate, wipe-on plate, pre-coated plate, multi-metal offset plate, deep-etch plate, static plate, dry planography, etc. Among them, the hand-drawn plate is the original lithographic platemaking method. It hand-delineates the printing part on the plate with automobile ink, which is no longer used now.

3. Classified by Printing Surface Texture

The surface texture of the printing plate is classified into the lithographic plate, dry-offset plate, and deep-etch plate.

II. Lithographic Platemaking Process Flow

The key technology of lithographic platemaking is to treat the surface of the printing plate into a blank part that is hydrophilic and a graphic part that is lipophilic. The lithographic platemaking process has developed into a very mature technology from the original technology based on the stone platemaking after continuous development. There are two processes: traditional Pre-Sensitized Plate (PS Plate) and computer-to-plate (CTP).

Section 2   Principle and Process of Traditional PS Plate Platemaking

I. Traditional PS Plate

PS Plate is an abbreviation for the Pre-Sensitized Plate.

The plate base of the PS plate is an aluminum plate with a thickness of 0.5 mm, 0.3 mm or 0.15 mm, etc. The aluminum plate is treated with electrolytic roughening, anode oxidization and hole sealing, etc., and then coated with a photosensitive layer on the plate to make a pre-coated plate.

II. Classification and Process of PS Plate

PS plate is classified into positive image PS plate and negative image PS plate according to the sensitivity principle of the photosensitive layer and platemaking process.

The platemaking process of the positive image PS plate is as follows:

Exposure → development → dirt removal → retouching → baking → development ink application → glue application

The exposure is that one side of the positive image negative film with the emulsion layer is stuck together with the photosensitive layer of the PS plate and placed in the special plate copying apparatus; after vacuum pumping is carried out, the light source of the plate copying apparatus is turned on to expose the printing sheet. The photosensitive layer of non-graphical part undergoes a light decomposition reaction under the illumination of light, whose commonly used source is the gallium iodine lamp.

Development is a development treatment of the exposed PS plate with a dilute base solution. It dissolves the compound generated by the light decomposition reaction, leaving a photosensitive layer on the page, which is not exposed to light, to form the graphic part of lipophilic. Development is generally performed by a dedicated developer.

Dirt removal is to remove the excessive regular line, sticking paper, the traces left on the edge of the negative film of the positive image, the dust and the dirt with the de-dire fluid.

Retouch is to add graphics to the developed PS plate for various reasons or to repair the plate. There are two commonly used methods, one is to apply a light-sensitive solution to the plate to make up for the graphic that needs to be added, and the other is to use the liquid supply pen.

Baking is to apply protection fluid to the surface of the printing plate after exposing, developing, de-dirt and repairing, and put it into the baking machine at the constant temperature of 230~250℃ for 5~8 minutes and then take out the printing plate; after naturally cooling, use the developing solution again to development; remove the remaining protection fluid from the plate and dry it with hot air. After the baking process, the PS plate's durability can be increased to more than 150000 prints. PS plate will not be baked if the number of prints is below 100000.

Development application is to coat development ink to a picture and text of the printing plate to increase their absorption to the ink, and to make it easy to check the quality of plate copying.

Glue application is the last process of PS plate printing, in which a layer of acacia is applied to the surface of the printing plate to make the hydrophobicity of the blank part of the non-graphic more stable and to protect the page from dirt.

The PS plate has fine sand meshes and high image resolution, and the resulting dot is smooth and intact, with a favorable tone and color reproduction.

For the PS plate that has been used in printing, erase the remaining ink and photosensitive layer on the plate, recoat the original aluminum plate base with the sensitivity solution to form a new photosensitive layer, then it can be reproduced for proofing or official printing. This method of remaking the PS plate by the used PS plate aluminum plate base is called the reproduction of the PS plate. It enables the aluminum plate base to be reused, so the PS plate is the most used printing plate in lithographic printing.

Section 3 Principle and Process of Lithographic CTP

I. Lithographic CTP

CTP technology emerged in the 1990s. 42 CTP systems were displayed at the Germany Drupa Printing Technology and Equipment Exhibition (Drupa) in 1995. At Drupa 2000, nearly 100 products were displayed by more than 90 computer direct-to-plate system and material manufacturers from all over the world.

CTP is a digital printing plate imaging technology. In CTP, a digital workflow is applied to transform words and images into digital directly, whose direct control is made by computer. Images are made by laser scanning, and then a printing plate can be directly printed by development. It eliminates the material of the film, the process of artificial imposition, and the semi-automatic or fully automatic plate copying process. The CTP technology, applying the CTP process (free of films), has many advantages including fast speed, high efficiency, fewer processes, simplified procedures, less variables, full-process digital processing, improved stability and standardized management.

CTP adopts a digital workflow to directly transform projects such as words and images into digital signals, to directly control the scanning imaging of the platemaking equipment with a computer, and then generate a printing plate through a processor. Fig.1-1 is the process flow chart of CTP.

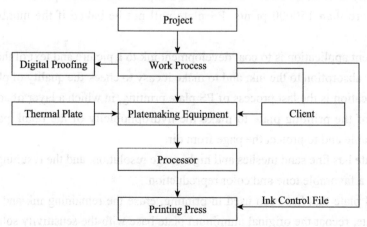

Fig.1-1   CTP Process Flow Chart

The structural principles of a platemaking machine are shown in Fig.1-2. It uses 16/32/48/64/96/ or 16 to 96 independent 830 /NM/1W or 405 /NM semiconductor lasers as the light source. N light sources are directed to the close-arranged plane through fiber coupling. The laser emitted from the close-arranged plane is focused on the plate closely and absorbed on the surface of the drum through the optical lens to perform thermal ablation or sensitising on the plate. The light drum equipped with the plate does high-speed rotation movements when it works. The scanning platform equipped with the fiber close-arranged system and the optical lens system does horizontal synchronization movements. The drive circuit system drives each independent laser to perform high-frequency switching according to the latex image of the computer, thus forming a latex image potential on the edition.

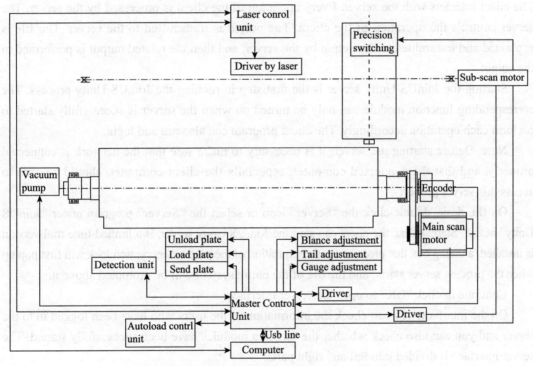

Fig.1-2　Structural Principle of Platemaking Machine

II. CTP of lithographic's Workflow

1. Workflow of Lithographic Platemaking

The work process of lithographic platemaking generally goes through five steps:

(1) Preparation before printing plate output: Set the parameters and template of printing plate through the output process software.

(2) Printing plate output: Realize laser exposure and output of photolithography plate through the operation of the platemaking machine.

(3) Development for printing: The exposed printing plate is passed through a developer so that the surface of the printing plate is treated as a flat printing plate that can be used on the machine.

(4) Printing quality detection: Observe the graphic of the printing plate output for platemaking, and test the relevant areas and labels to judge the printing quality of the printing plate.

(5) Post-processing of printing plate: In order to improve the printing adaptability of the printing plates, the exposed and developed printing plates also need to go through post-processing procedures such as baking and gluing.

2. Preparation before Printing Output (exampled by JoinUS Unity software)

(1) Steps of the Process Software

① Start the JoinUS Unity Server

JoinUS Unity is based on the C/S work model. The server is the core of the entire process. The client interacts with the server. Every operation of the client is processed by the server. The server controls the operation of the client. The output is transmitted to the server. The file is explained and transmitted to the output by the server, and then the related output is performed in the output.

Starting the JoinUS Unity server is the first step in running the JoinUS Unity process. The corresponding function module can only be turned on when the server is successfully started to perform each operation accordingly. The client program can also run and login.

Note: Before starting the server, it is necessary to make sure that the network is connected properly, and that the connected computer, especially the client computer, should be able to access this server properly.

On the desk, double-click the "Server" icon or select the "Server" program under "JoinUS Unity" in the "Start" menu, the client can start the JoinUS Unity server. If a limited-time trial version is installed, a dialog box that prompts for the expiration of the process encryption lock will first pop up when the process server starts, and the life of the entire system will be prompted above it.

Continue to click "OK" to open the server interface.

On the interface, you can check the information of the users who have been logged in to the server and you can also check whether the function modules have been successfully started. The server interface is divided into left and right parts.

Connection information of the client: The information of the client logging into the server is shown on the left, including the ID, username, IP address, etc.

ID: Login number, which is generated automatically according to the login order;

Username: The name of the user who is logging in or the name of the output;

IP address: The IP address of the host where the client or output is logged;

Information of work order processor: Function modules that have been started by the server are shown on the right.

Successful connection to the database: It indicates that the database is successfully started and connected to the server properly. The process will not work properly if "Database Connection Failure" is shown.

Normalization module started: It indicates that the normalized processor module has been successfully started;

Imposition module started: It indicates that the imposition processor module has been successfully started;

Output module started: It indicates that the output processor module has been successfully started.

"Encryption Dogs Error!" will be prompted if encryption dog are not detected at runtime or if the authentication code used by the software does not correspond to the currently used encryption dog.

② Launch the JoinUS Unity Client

The client is user's operation port, in which the user submits the operation of normalization, imposition, output, etc. to the server, and displays the results of each step of processing in the client.

Users can only start the client after the process server has been successfully started.

When logging into the client, initially, users should make sure that the connection between the machine and the server is normal, that the client and server should be set up in a same local area network, that each machine has its own IP address so the client can gain better access to the server. Secondly, only one client can be started on one machine, i. e., multiple clients are not allowed to operate on one machine at the same time. users cannot log in with the same username repeatedly when they start the client on different machines, i. e., they can only log in once for the same user.

On the desk, double-click the "Client" icon or choose "Client" under "JoinUs Unity" on the "Start" menu. The client login dialog box can be opened later, as shown in Fig.1-3.

Fig.1-3 Client Login Dialog Box

In the dialog box, enter the correct username, password and server IP address, then click "OK" to open the JoinUS Unity client. If the IP address entered is not right, you will be prompted to "fail to connect to the service" and click "OK" to exit the login interface. After the JoinUS Unity client has been logged in, the login information of the user will be displayed on the Unity Server interface.

③ Start the JoinUS Unity Output Unit

The JoinUS Unity output unit is an important part of the process, which is responsible for the output of the entire process, receiving the page files parsed by the server, and controlling the output of the files to the corresponding equipment.

The same type of output in the process is managed according to the IP address, so only one output can be started on each computer. The output can be started on the same computer as the server.

> Computer to Plate of Lithographic

  The JoinUS Unity output is divided into CTP output and digital printing output. Double-click the corresponding JoinUS Unity output on the desktop or click "Start" → "Programs" → "JoinUS Unity" → "CTP Output or Digital Printing Output" to open the output boot interface.

  The appropriate Output Login dialog box can be opened later. Enter the correct server IP address in the dialog box, which is the IP address of the computer where the server is located. In this way, the server can be successfully connected. Click the "OK" button to open the corresponding output interface.

  Before starting, make sure that both computers are accessible.

  After the output is successfully logged in the information of logging into the output will be displayed in the "Process Server" interface. It indicates that the corresponding output has successfully signed into the server.

  (2) Work Process of the Process Software

  ① Create a Project

  First, a new project is required, and a unique job ticket number should be entered for each project. After the creation of the project is completed, a corresponding project folder will be generated in the file system of the process server, as shown in Fig.1-4.

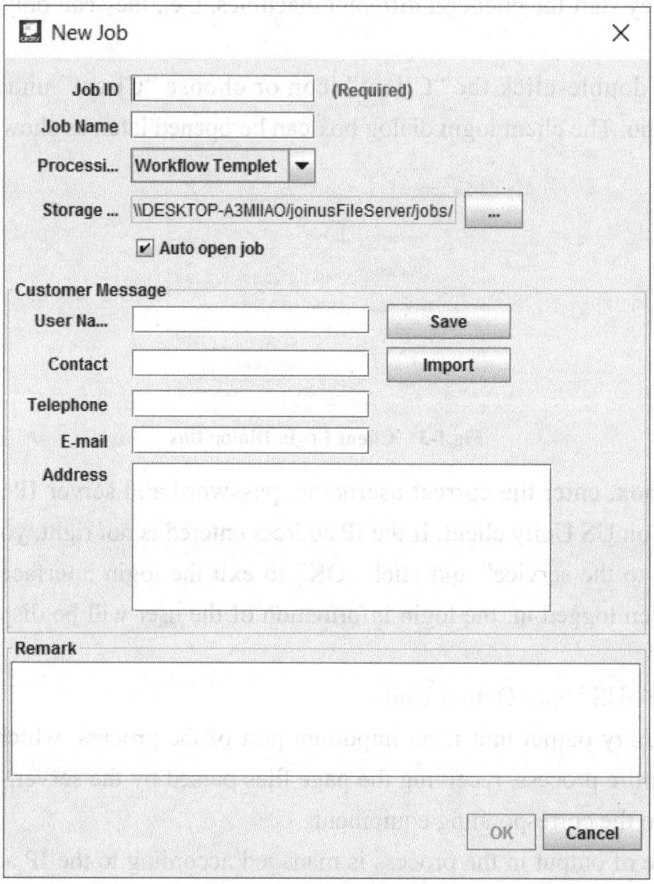

Fig.1-4 New Project

Many process templates can be added to the project as required. Usually, at least one normalized template is required and the template should be set up relatively. These process templates, which include JTP parameters, constitute the "digital job ID" for this project.

Job ID: The job ID is the unique identification of projects in the whole system and cannot be repeated with that of other projects;

Job name: The name of the project.

Process template: Select a process template defined in the template library for the project;

Customer information: The information and job ID of the customer to which this project belongs is required. The project name can be the same as the job ID or you can name it differently. None of the names can contain the following characters:/? \ | * : " "<>。

② Set up a Project Process Template

The main function of the "Process Template" interface is to create a process template required for the current project. Users can create required templates according to the needs of this project. A process template will be shown in the process template interface if it is preset during a new project, as shown in Fig.1-5.

A template is a parameter setting data of each processing node of a process. Each node means a specific operation in the process. Each node template required to process this project is saved in the project process templates. A node template needs to be added to the project firstly, if users want to use a processing node.

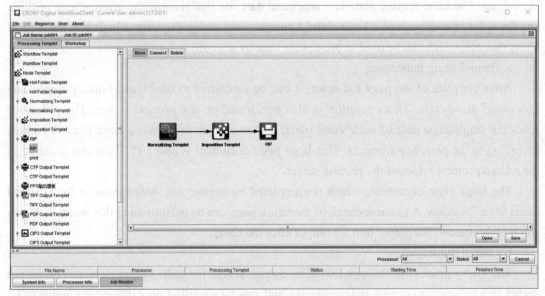

Fig.1-5　Processing of Open Project Template

a. Add source file

After setting up the process template, users should add source files to the project. The process uploads the added source files to the project folder on the server. These source files is also listed in the source file area of the workshop.

b. Normalized documents

The user submits the source file to the normalized template and triggers the normalized processor on the process server to start work. The JTP reads the corresponding source file and conducts operation processing according to the parameters in the normalized template to generate a normalized PDF file. The PDF file after the specification exists as a single page, known as "the Normative Page". The normative page data is also saved in the project folder on the process server.

Note: From the above description, it can be found that during the normalization process, the client only issues instructions, and all execution is performed in the process server, i.e., the CPU of the server executes the normalization program, reads the source file from the server and processes it, and stores the processing results back to the server. Throughout the process, no page data is passed between the server and the client workstation, which is a typical mode of operation for a client/server system.

c. Create a list of pages

After being normalized, the specification pages can be sorted into a page list in the Page List window, which can be organized into a page list by the normative pages of different source files. A list of each page can be regarded as a book with a page number order.

③ Layout

For a collated list of pages, it needs to be given a hand-folding plate for hand-folding imposition. The hand-folding plate is a seperated data file that records the arrangement position and the order of each page on a large plate. A special plate design program is built into the JoinUS Unity process, which can be used to pre-design various imposition formats.

a. Hand-folding imposition

After the plate of the page list is set, it can be submitted to the "Hand-Folding Imposition Processor" to operate. This operation is also performed on the process server. The processor reads the single-page data of each small plate, and assembles them into a large plate document according to the plate requirements. This large plate document is also a PDF file and is stored in the related projects folder of the process server.

The large plate document, which is completed by imposition, will appear in the "Printed Sheet Style" window. A preview check of the entire page can be performed in this window, which can be submitted to the output unit for output after the check.

b. Output printing plate

The Unity process is equipped with an output server for each output device. The output server is a program that runs independently and can be installed on different computers. An output template for such devices needs to be added in the process project in order to use one of the output servers. The IP address in the template is pointed to the computer where the corresponding output server program is located. Then, users submit the finished plate of the large plate document to this output template. The file is exported to the specified follower. When the output page appears in the project list of the follower, all prepress processing on this page, including rasterization and screening, has been performed. The last dot preview inspection can also be performed in the output server, including dot preview and overprinting inspection. After the inspection, it can be output to the printing plate for printing or output to a digital printing machine to print the finished product.

3. Plate Output

The entire workflow of CTP is shown in Fig.1-6.

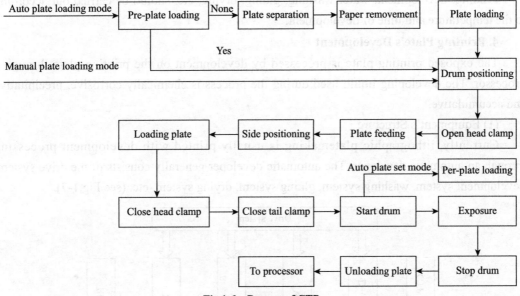

Fig.1-6 Process of CTP

Template establishment: parameter setting (page frame, accuracy, exposure power).

Plate set: In the case of automatic plate set mode, the first judgement is whether the plate set has been pre-formed. If there is no pre-plate set, move the sucking mouth down and judge whether it is the paper or plate. If it is paper, remove the paper and return to perform the previous action; if it is the plate, the plate set will go to the entry port of the hosts. In case of manual mode or plate set, skip the previous action.

Plate load: Position the light drum first, and then open the head clamp after the light drum is positioned at the plate load. The plate feed system then delivers the plate to the head clamp for front positioning. The side gauge system locates the sides of the sheet. Then close the head clamp. Then press plate load roller down. The light drum will turn to perform plate load. Turn on the tail clamps when it is turned to the tail, the light drum will be reversed; the tail will be sent under the tail clamp, and the tail clamp will be closed. The plate load roller will be up.

Imaging: Start the light drum. In the case of automatic plate set mode, start the plate setter again at the same time. The exposure begins when the light drum's speed reaches the set value, and the vacuum pressure is achieved. The laser system on the optical platform illuminates the laser beam carrying the computer latex information on the plate to realize the transmission of word and image information. The optical platform moves in a straight line at a constant speed. The roller rotates for one circle. The platform moves corresponding to the line distance of the optical path. The end of the dot matrix information means that the first plate of the exposure is completed. After that, the light drum decelerates until it stops.

Plate unload: Position again after the light drum stops. Plate unload mechanism starts to unload plate. The typesetting mechanism delivers the plate into the bridge. The plate goes through the bridge to a processor.

Printing: development, water washing, gluing, drying, etc. should be carried out according to the temperature and time of development.

4. Printing Plate's Development

The exposed printing plate is processed by development on the printing plate through a processor. The developing liquid used during the process is chemically corrosive, precipitative and accumulative.

(1) Equipment's Structure

Currently, lithographic platemaking is usually printed with development processing through an automatic developer. The automatic developer generally consists of the drive system, development system, washing system, gluing system, drying system, etc. (see Fig.1-7).

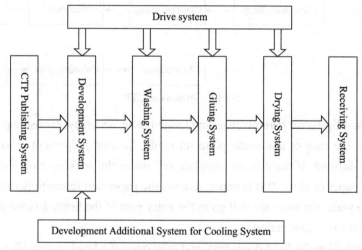

Fig.1-7 Processor's Structure

① Drive System

The drive system is a drive device that guides the operation of the printing board. The motor drives the conveyor roller through the worm or sprocket chain, and the printing plate is driven through each working link of the developer by the pair of press rollers.

② Development System

During the printing plate development, the developing tank ensures that the developing liquid can work in a constant condition through the heater, cooling system and circulating filter system. The brush roller scours the printing plate under certain pressure and speed when the printing plate passing through the developing tank. The photosensitive layer on the printing plate dissolves rapidly.

③ Washing System

The washing system consists of two sets of spray pipes to spray clear water to the front and back of the printing plate to remove the development product and the residue developing liquid, and squeeze out the water from the plate by extrusion roller. The system has the function of secondary water washing.

④ Gluing System

The gluing system is a thin layer of protective glue applied to the cleaned printing board to ensure that the printing plate is cleaned and antioxidant before printing.

⑤ Drying System

The drying system consists of an air supply pipe and cone. The printing plate is dried rapidly by air blowing and heating.

(2) Commissioning Operation Instructions [taking Huqiu Imaging (Suzhou) Co., Ltd. PT series printing processor as an example]

Enter the commissioning stage of the machine after the installation is completed. First, fill the development liquor tank with clear water until it spillovers from the spillover port (if the machine is equipped with an inner cycle of water washing, clean water should also be added to the washing tank until it spillovers), and check whether the machine is in a horizontal position. Development replenishment fluid and glue can be replaced by clear water first.

Note: Avoid splashing water into the electrical components while infusing clean water; dry it with a hairdryer if needed. Otherwise, the test mustn't be powered on.

Power on and turn it on. First, set the necessary process parameters, such as development temperature, motor speed, plate size & specifications, etc. by using the operation interface. After the process parameter setting is finished, select the corresponding program to put the machine in operation. The machine enters the heating state and then observes whether the circulating pump ground is working properly at the same time. Details can be found on the flow state of the liquid or using a thermometer to measure whether the temperature of development is rising. It is recommended that users manually test the washing functions before official debugging and observe whether the functions are normal after the official operation.

The pressure of each speed shaft has a direct impact on the quality of publishing, so it is crucial to adjust the pressure of each speed shaft. However, due to the great differences in the plate materials and developing liquid used, the pressure requirements for each speed shaft are also different. Therefore, refer to the pressure range in Table 1-1, and adjust the pressure of each speed shaft to the appropriate range to meet the actual publishing needs according to the use of the plate materials and developing liquid.

Measure method: Cutting a 30 mm wide plate from a plate with a thickness of 0.27 mm, and fixing one end of the tension gauge with the other end of the plate by a suitable method. The other end of the plate is then inserted from the middle of a pair of shafts to be tested to measure the pulling force of the roll shafts.

Table 1-1 Processor's Parameters

Shaft Serial Number	Pull	Remarks
Development 1 (rubber shaft)	60~70 N	
Development 2 (brush shaft)	3.7~3.8 N	
Development 3 (brush shaft)	3.7~3.8 N	Test environment: Temperature 28℃ Humidity 50% Specifications of the plate: 30 mm in width 0.27 mm thick
Development 4 (rubber shaft)	80~90 N	
Wash 1 (rubber shaft)	60~70 N	
Wash 2 (brush shaft)	3.5 N	
Wash 3 (rubber shaft)	80~90 N	
Glue 1 (rubber shaft)	80~90 N	

Note: The data in the table are all measured when the surface of the axis is dry.

Turn off the processor's main power supply after the test is completed. Drain the clean water in the processor (water in the filter should also be drained). After draining the water, don't forget to reset the adjusted valves. Then inject the developing liquid into the developing tank until it overruns. If the water washing internal circulation function is equipped and it is required, inject clear water into the washing tank until it overruns. Replace the clear water in the development replenishment drum and the glue drum with development replenishment fluid and glue respectively.

Power on, formal processing test. Set the process parameters (such as development temperature, development time, brush speed, etc.) according to the requirements of the plate materials and developing liquid client actually used. After the machine meets the requirements of processing, the official printing will be carried out. Fine adjustments are made to the related functions of the processor (development time, reel axis pressure, sizing volume, etc.) according to the published test results to make the washing effect of the plate meet the requirements.

(3) Announcements

Read the main parameters and technical indicators of the product carefully and confirm whether the user processing requirements are consistent with the equipment indicators.

Attention should be paid to whether the equipment is operating properly after starting up every day.

The flow of washing water should be observed regularly, especially in areas where the water pressure is unstable. Fit the flow pressure gauge if necessary. The drain should be done at the end of the work every day.

Do not put the plate materials whose width and length do not meet the specifications listed in the table nor rolled plate into the washing machine. Press the stop button to check immediately to avoid the plate being stuck once the plate materials fails to come out from the publishing port within the set time.

Clean and maintenance must be carried out regularly according to the maintenance methods provided.

It is advisable to turn it off on standby when finishing processing every day. Turn off the power switch and the air switch in turn, and finally turn off the main power switch and the faucet. Never pull the power switch directly, which will affect the service life of the machine.

It is strictly forbidden for non-professional personnel to open the electrical box without permission, so as to avoid personal danger and equipment malfunction.

5. Post-processing of Printing Plates

The printing plate after development processing needs to be inspected and tested. After confirming that the printing plate is free from errors and the quality is in line with the standard, post-processing is required for the printing plate. Currently, many processors can directly post-process printing plates, i.e., by gluing and drying systems.

(1) Baking Plate

The baking plate is to apply the protection solution to the surface of the printing plate after exposing, development, dirt removal, and repair, then put it into the baking machine; bake for 5~8 minutes at a constant temperature of 230~250℃, and take out the printing plate; after naturally cooling, use developing liquid again for development; remove the remaining protection solution from the plate and dry it with hot air. Durability can be improved after the baking process for the printing plate; however, the printing plate does not need to be baked if the number of prints is not high.

Note:

① For the PS plate, users must remove the tape and dirt points on the page between the baking plates because the photoconductor after the baking plate is adsorbed on the page firmly.

② The protective glue of baking plate should not be erased too much, otherwise, traces will easily appear and in severe cases, it is not easy to get ink.

③ Rinse the retouching solution with clear water after retouching. Otherwise, the remaining retouching solution and the dissolved substance will contaminate the plate and cause dirt after the baking.

④ The baking plate can only be performed after the protective glue is dried.

⑤ The universe protective glue of the baking plate should be done with degreasing gauze to avoid using a dirty cloth to contaminate the plate and cause dirt.

⑥ Avoid excessive force in applying the protective glue for the baking plate, so as not to affect the quality of the baking plate by shedding fibers.

(2) Gluing

Glue on the printing plate can protect the small sand object on the surface of the plate base, enhance the wear resistance of the sand object and improve its durability; it can protect the hydrophilic layer of the plate base, improve the inking property of the graphic part, which is conducive to the quick water-ink balance during printing; it also can protect the printing plate to avoid slight scratches; it can effectively seal the sand object from dust waiting in to prevent the printing plate from being directly exposed to the air and causing excessive oxidation.

Gluing is the last process of lithographic platemaking, i. e., applying a layer of acacia gum to the surface of the printing plate to make the hydrophobicity of the blank part of non-graphic more stable and protect the plate from dirt.

Chapter 2 Basic Knowledge of Lithographic Platemaking

Objectives:

1. Basic working principles of CTP machine;
2. The output resolution and its setting;
3. Role of screening;
4. Master the format and characteristics of lithographic platemaking ;
5. Understand the lithographic platemaking backend font;
6. Master the basic theory of page printing
7. Understand the dot technology of CTP screening;
8. Master the theories related to the compensation curve of the printing press.

Section 1 Basic Working Principles of CTP Machine

I. CTP Machine and Its Structure

CTP is a digital printing plate imaging technology. The CTP machine consists of three parts: an accurate and complex optical system, a mechanical system, and a circuit system.

The optical system consists of a laser, fiber coupler, close-arranged head, optical lens, optical energy measurement, etc.

Machanical system: machine frame, light drum, wallboard, plate sending department, head opening/closing part, plate load roller, plate unload department, typesetting department, screw guide track, scanning platform, etc.

Circuit system: encoder, main/auxiliary servo motor, each actuator stepping motor, vacuum

Chapter 2 Basic Knowledge of Lithographic Platemaking

pump, main control board, terminal board, laser drive plate, position sensors, etc.

A single beam of original laser generated by the laser machine is divided into multiple beams (usually 200~500 beams) of extremely fine laser beams through multiple fibers or complex high-speed rotation optical beam splitting systems. Each beam of light is characterized by the bright and dim of the image information in the computer through the acoustic-optical modulator, respectively. The bright and dim changes of the laser beam are modulated and turned into a controlled beam. After focusing, hundreds of micro laser beams shoot on the surface of the printing plate for engraving. A shadow of the image is formed on the printing plate after scanning the engraving. After development, the image information on the computer screen is restored to the printing plate for direct printing by the offset press.

The CTP machine consists of three parts: an accurate and complex optical system, a circuit system and a mechanical system.

II. Variety and Imaging Principle of CTP Machine

The CTP system can be divided into inner drum type, outer drum type, and flat plate type from the perspective of the exposure system. In terms of plate materials, it can be classified into silver salt edition, heat-sensitive edition (ablating heat-sensitive edition, non-ablating heat-sensitive edition), photo-sensitive resin edition and polyester edition (non-metallic plate base), etc.; in terms of technology, it can be classified into heat-sensitive technology (ordinary laser imaging), purple laser technology, UV light source technology, etc..

1. Diazo Plate

The diazo plate is usually used as a flat printing plate. The number of photochemical molecules in the diazo plate is directly related to the absorption of light quantity and overlap, and the sensitivity range is limited to the ultra-violet region of the spectrum and is free from expansion effect. It is suitable for exposure of high-power Artemisia lasers.

2. Photopolymer-type Plate

The performance of photopolymer-type plates is similar to that of diazonium type. And there is no expansion effect. Many types of photopolymers are used. Its mechanism is similar to that of a diazonium plate and is used for preparing flat printing plates. The cross-linked image is formed after exposure and is suitable for exposure by an ultra-violet laser.

3. Silver Salt Light-sensitive Plate

The mechanism of silver salt photoactive plate materials is the use of photochemical reaction to transform silver salt into metallic silver and the catalytic expansion of these silver particles to promote the reduction of silver ions around them.

4. Photosensitive Resist Plate

Like photopolymer and diazo plates, the sensitivity range of photosensitive resist plate is usually limited to the ultra-violet portion. The photoresist, MC929, produced by KODA, is used in a flat concave magnesium plate and can be exposed to high-quality images using an Artemisia laser and a visual laser.

Section 2 Output Resolution and Its Setting

1. Device Pixels

The printing plate is exposed to laser during production to form lines, text and dots on the printing plate. Each of these exposure points is a device pixel.

2. Output Resolution

The number of output device pixels in a unit of distance for printing plate is called "output resolution", in dots per inch (dpi).

The denser the device pixels are, the finer the image will be printed. In other words, the higher the output resolution, the finer the output.

2400 dpi is the default output resolution of the output system, which can meet the needs of fine printing. The curve output at 2400 dpi is very smooth and the small size words are also very clear.

3. The Relationship between Output Resolution and Screen Frequency

Screen frequency (lpi) is only the resolution of the screening part, while the output resolution (dpi) is the resolution of the whole printing sheet.

Different screen frequency can be found on a printing board, for example, 175 lpi can be used as a whole on a printing board, and 100 lpi is required for a picture for special effects. However, the same output resolution is used for the entire printing plate when output.

The screen frequency is much smaller than the output resolution. For color printing, their configuration is usually 175 lpi and 2400 dpi.

Increasing the screen frequency to a certain extent enables a more refined part of the screening, but this does not affect the un-screening solid, text, and lines. Improvement in output resolution not only refines the concepts of field, text and lines but also makes the dot plumper, by which we improve the output quality of the screening section. Therefore, pure text printing requires sufficient output resolution rather than screen frequency. A printed article with both pictures and text requires both. For example, the screen frequency of a text-only novel can be 80 lpi. But with illustrations, it should be at least 133 lpi, but output resolution for both is 2400. The output company sets the output resolution to 2400 dpi for various types of businesses uniformly. The relationship between screen frequency and output resolution is

$$\text{Screen frequency (lpi)} = \frac{\text{Output Resolution (dpi)}}{\text{Width of Mesh Unit (Tone, Device Pixel)}} \quad (2\text{-}1)$$

For example, if the output resolution is 2400 dpi, and the width of a mesh unit is 16 points (i.e., 16 device pixels), then the screen frequency is 2400/16=150 lpi.

In the case of fixed output resolution, increasing screen frequency can only reduce the number of device pixels that can be accommodated by the mesh modulation unit. Namely:

$$\text{Width of Mesh Unit (tone)} = \frac{\text{Output Resolution (dpi)}}{\text{Screen frequency (lpi)}} \quad (2\text{-}2)$$

4. Relationship between Output Resolution and Image Resolution

Three cases are required to consider for the relationship between output resolution (dpi) and image resolution (ppi) as follows:

(1) For all screening images, according to the screen frequency setting the image resolution only (usually 300 ppi), output resolution must be considered by the designers.

(2) For all unnetted images, the closer the image resolution to output resolution, the higher the printing quality. Usually, it is 2400 dpi at the output. If the resolution of the image is set to 2400 dpi, the best printing effect will be obtained (only in this case that output resolution and image resolution are equal, and a bit pixel corresponds to a device pixel). However, 900 dpi, 1200 dpi or other dpi can also be picked if 2400 dpi makes the image file too large.

(3) For some screening images, only 300 ppi is required for the screening part, but more than 900 ppi is required for the unscreened section. Full care for the unnetted part may make the file too large, where an intermediate value, such as 600 dpi, can be selected.

Section 3 Role of Screening and Its Setting

I. Dot

The printing uses dots to reproduce the original manuscript. When the printed product is enlarged, it will be found that its graphics composed of countless dots of different sizes. The larger the dot, the darker the color will be printed; vice versa.

The arrangement position and size of the dot is determined by the netting number of lines, for example, the number of points of screening is 150 lpi means that 150 dots are along the length or width of 1 inch. The dot of different colors will be staggered at different angles to avoid overprinting all colors of ink.

II. Screening

In prepress processing, the screening method is used. The dot presents a change in light and dark with a certain number of pixel combinations. A dot can consist of pixels in arrays of different sizes.

Currently, the screening method mainly includes Amplitude Modulation (AM) screening and Frequency Modulation (FM) screening. Most lithographic printing processes use AM screening, and AM screening is controlled by three screening parameters nowadays.

(1) Dot Shape

The shape of the dot in printing is divided by the geometric shape of the dot in the case of a 50% inking rate, which can be divided into square dots, circular dots, and diamond-shaped dots. Fig.2-1 shows the shape of the printing dot presented, under the microscope.

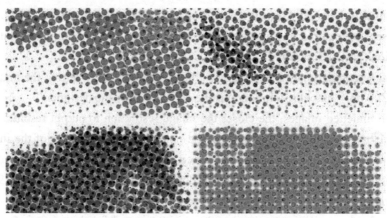

Fig.2-1   Printing Dot Presented under a Microscope

The square dot is checkerboard shaped at 50% coverage. It is sharper in particles and has a strong ability to express on levels. It works well with the performance of lines, graphics and some hard-toned images.

The circular dot is independent of the dot in both the light and the case of intermediate tone. It is only partially connected in the case of dark tone. Therefore, it is less common for four-color printing because the poorer performance at the tone level, as shown in Fig.2-1.

The diamond-shaped dot combines the hard tones of the square dot with the soft tones of the circular tone. It has a natural color transition and is suitable for general images and photos.

2. Dot angle (see Fig.2-2)

In printing platemaking, the choice of dot angle plays a vital role. When the wrong dot angle is selected, Moore streaks will occur.

Common dot angles are 90°, 15°, 45°, 75° and so on. The dot of 45° performs the best and is stable without appearing rigid; the angular stability of 15° and 75° performs not that well but the visual effect is not rigid; the angle of 90° is the most stable, but the visual effect is too rigid to be beautiful.

Two or more dot types are overprinted at a certain angle, which will produce a certain light transmission and shading effect. Moore streaks will be generated when the angle is small, which will seriously affect the beauty and quality of the printed image, which is commonly known as "Moire".

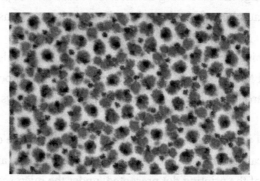

Fig.2-2   Round Dot with Degrees

> Chapter 2 Basic Knowledge of Lithographic Platemaking

In general, the overall streaks are more attractive when the angle differences between the two-tone types are 30° and 60°; the angle differences of the dot types are 45° can have a good performance, too; and the streaks are generated when the angle differences of the two-tone types are 15° and 75° will affect the quality of the printed image.

3. Number of dot lines

The size of dot number of lines determines the fineness of the image, as shown in Fig.2-3. The common dot number of lines are as follows.

10 ~ 120 Lines: For low-quality printing, large areas of prints such as posters and placards which are usually viewed from a long distance and are generally printed using newsprints and offset paper; sometimes low-gram sub-powder paper and coated paper are also used.

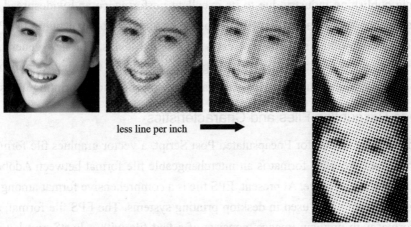

less line per inch →

Fig.2-3 Screen Frequency and Image Fine Degree

150 Lines: This accuracy is generally used in ordinary four-color printing, available on all types of paper.

175 ~ 200 Lines: This is suitable for exquisite picture books, pictorials, etc., most of which are printed on painted paper.

250 ~ 300 Lines: For paintings or other products with the highest quality requirements, most of which are printed with advanced paint paper and special paper.

Section 4   Lithographic Platemaking Output File Format

I. Page Description Language

Page description languages have developed rapidly and have been widely used since their birth in the 1980s. The page description languages mainly include Post Script by Adobe Systems,

Interpress by Xerox, DDL by Image, PCL 5 by HP, etc. The most famous and widely used one is the PostScript language (PS language).

PS Language was born in 1976 at Evans & Sutherland, a computer company in the United States. It was designed as a page description language for electronic printing. It had been modified serveral times. It was not officially named until Adobe Systems, founded by Charles Geschke and John Warnock in 1982, realized this language again. It launched the first Apple laser printer equipped with a PS language interpreter in 1985. After just a few years, the PS language has been widely used and has become an indispensable part of high-quality graphic printing and output. In 1990, Adobe developed the more powerful PostScript Level 2 for color files and Display PostScript (display version PS), for workstation multi-window environments. It has become the defacto industrial standard for the electronic publishing industry due to its excellent performance and widespread application of the PS language. Adobe has improved the PostScript standard by expanding the PS language to provide complete printing solutions for users at different levels. The new generation of page description languages is available in a wide range of applications, from homes and small offices to group companies, from printing device manufacturers to professional printing industries.

II. Page Description Files and Characteristics

EPS is the abbreviation for Encapsulated Post Script, a vector graphics file format based on the PostScript language. EPS format is an interchangeable file format between Adobe Illustrator and Adobe Photoshop software. At present, EPS file is a comprehensive format among the generic exchange formats commonly used in desktop printing systems. The EPS file format, also known as the PS format with preview images, consists of a text file with a PostScript language and a (optional) low-resolution representative image described in PICT or TIFF format. An EPS file is a PostScript file that includes file header information, which can be used by other applications to embed this file in a document.

The "encapsulation" unit of the EPS file format is a page. In addition, the page size can be determined by the overall boundaries of the objects on the saved pages. Therefore, it can be used to save both a standard page size in the software version and a rectangular area of an object of independent size.

Its text part can also be written by ASCII characters (so that the generated file is large but can be modified and examined directly in a normal editorial) or by binary digits (small files are generated, which are fast to process, but are inconvenient to modify and check).

The EPS file can also accommodate latex images, although it uses a vector description. It just does not transform a lattice-matrix image into a heddle description but saves all pixel data as a whole in a pixel file. For the grouping and output control parameters for pixel images, such as parameters of contour curve, screening parameters and dot shape, color device property file (Profile) of images and color blocks, etc., they are saved separately in PostScript language.

The EPS file can be in multiple forms, such as CMYK EPS in color space (including a PostScript description of the four-color separation image and an optional low-resolution representative image), RGB EPS, and L^*a^*b EPS. In addition, different EPS files generated by different software are also different, such as Photoshop EPS, Generic EPS, AI (Illustrator software version in EPS format), etc. Attention should be paid to compatibility in cross-use.

EPS format supports embedding color information ICC property file in the file (see Fig.2-4). The EPS file can be embedded in two color information files: proofreading setting information and ICC property file. When embedded proofing setting color information is selected, the color information of the image file will be transformed according to the information in the proofing setting, that is, the color information on all images will be transformed according to the characteristics of the proofing setting. When the ICC property file is selected, the color information of the image file is only with the information of the output state. The color information of the image will not change in any way.

PostScript color management converts file data to the color space of the printer. Do not select this option if you plan to put the image in another document with color management. Only the PostScript Level 3 printer supports PostScript color management for CMYK images. To print a CMYK image using PostScript color management on the Level 2 printer, convert the image to Lab mode and then store it in EPS format.

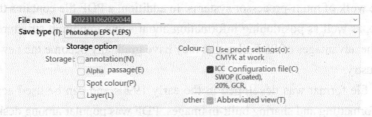

Fig.2-4 Photoshop Storage in EPS Format

EPS files can also carry all information about font libraries related to words. It should be noted that only low-resolution representative images can be output when exported to non-PostScript devices. A high-resolution output can only be obtained on the PostScript output device. So, in many cases, the proofs operator printed are very rough. The reason is that operators are using a non-PostScript printer to print PostScript files. If it is replaced by a PostScript printer, a nearly perfect graphic will be printed.

III. Portable Document Format (PDF) and Its Characteristics

Portable Document Format, also known as PDF, is a file format developed by Adobe Systems for file exchange that is independent of applications, operating systems, and hardware. PDF files are based on the PostScript language image model. Accurate colors and printing effects are ensured on any printer, which means that the PDF files will faithfully reproduce every character, color and image of the original manuscript.

The Portable Document Format is an electronic file format. This file format is independent of the operating system platform. In other words, PDF files are generic in the Windows, Unix, or MacOS of Apple company. This makes it an ideal document format for electronic document distribution and digital information dissemination on the Internet. PDF files are more and more common in e-books, product descriptions, company posters, network materials, and emails.

Adobe designed a PDF file format to support cross-platform cooperation, multimedia-integrated information's publishing and releasing, especially to provide support for network information releasing. For this purpose, PDF has many advantages that other electronic document formats can not compare to. The PDF file format enables text, glyph, format, color, and graphical images that are independent of the device and resolution to be encapsulated in a single file. It can also contain electronic information such as hypertext links, sounds and dynamic images, and supports special files with high integration and security reliability.

The PDF consists of three technologies.

(1) Derived from PostScript: Used to generate and output graphs;

(2) A glyph embedded system: It enables the glyph to be transmitted with the file.

(3) Structured storage system: With the appropriate data compaction system, it is used to bind the elements and any related content to a single file.

PDF files use an industry-standard compaction algorithm, which is usually smaller than PostScript files and easy to transmit and store. It is still page-independent. A PDF file contains one or more "pages", which can be processed independently for pages, which is especially suitable for the work of multi-processor systems. In addition, a PDF file contains the PDF version used in the file, as well as positioning information about some important structures in the file. It is because of the advantages of PDF files that they have gradually become the new favorite in the publishing industry.

The PDF file format was developed in the early 1990s and can be used across platforms, including file formatting and sharing built-in images. PDF was popular among desktop publishing workflow technology in the first few years when the World Wide Web and HTML text had not yet emerged.

Starting in version 2.0, Adobe started to distribute Acrobat Reader for free, a reading software for PDF (now renamed Adobe Reader, which is still called Adobe Acrobat), while the old format could still be used, making PDF a later informal standard for the fixed format text industry. As of 2008, Adobe Systems' PDF reference version 1.7 became ISO 32000-1:2008. Since then, PDF has become an official international standard. It is also for this reason that the current development of updated versions of PDF (including the development of future PDF 2.0 version) has become dominated by TC 171 SC 2 WG 8, but experts from Adobe and other related projects are still involved.

Initially, the PDF was only considered as a page preview format, not a production format. Adobe announced on July 13th, 2009, that PDF/Archive (PDF/A), as a long-term preservation format for electronic documents, has been approved by the Standardization Administration of the P.R.C. as a Chinese national standard, and has been officially implemented since September 1, 2009. PDF files have become a defacto industry standard for digital information.

1. ISO Standardization

Since 1995, Adobe has participated in some workgroups that created publishing technical specifications by ISO and collaborated with ISO in the process of a subset of PDF standard specialties (such as PDF/X or PDF/A) for specific industries and applications. A subset of the full

PDF specification is developed to remove those are not required or cause problems for a specific purpose, as well as the use of some required functions that are just optional (not compulsory) functions in the full PDF specification.

On January 29, 2007, Adobe announced that it would release the full PDF 1.7 specifications to the American National Standards Institute (ANSI) and the Association for Information and Image Management (AIIM). The International Organization for Standardization (ISO) will release the full PDF 1.7. Adobe will develop PDF specifications as a future edition, and Adobe is just a member of the Technical Committee of ISO.

The standard for ISO "Full-featured PDF" is published under official No. ISO 32000. A fully functional PDF specification means more than a subset of Adobe PDF specifications. For ISO 32000-1, the full-function PDF contains each of the PDF 1.7 specification definitions of Adobe. However, Adobe later released an expansion that is not part of the ISO standard. Those are also proprietary functions in PDF specifications, which can only be used as additional specifications.

2. Version History

(1) PDF 1.0

PDF 1.0 was released in the fall of 1992 in COMputer DEaler's eXpo (COMDEX), and the technology won the Best of COMDEX. Acrobat, a tool for creating and viewing PDF archives, was launched on June 15, 1993. It already has internal linkage, bookmark and embedded font functions, but the only color space supported is RGB, which is not supported for prepress operations.

(2) PDF 1.1

Acrobat 2, which was launched in September 1994 and supports the new PDF 1.1 archive format. New features of PDF 1.1 include External Link, Article Threads, Security Features, Device Independent Color and Notes.

(3) PDF 1.2

In 1996, Adobe launched Acrobat 3.0 and supported PDF 1.2 specifications. PDF 1.2 is the first PDF version that is truly available in the prepress environment. In addition to forms, it includes support for OPI 1.3 specifications, support for CMYK color space, and prepress-related functions such as special color definition, Halftone function and Overprint in PDF.

(4) PDF 1.3

Adobe launched Acrobat 4, known internally as Stout, in April 1999, which brought us PDF 1.3. The new PDF format specifications include support: two-bit CID fonts; OPI 2.0 specifications; a new color space called DeviceN that improves the support for special color capabilities; smoothing shading, a technology that is efficient and very smooth to step down from one color to anothor, annotations.

(5) PDF 1.4

Adobe launched Illustrator 9 in mid-2000, which is indeed an amazing characteristic: it is the first application to support PDF 1.4 and its transparency characteristics. It is the first time that Adobe has introduced a new version of Acrobat not along with the new version of PDF specifications. Nor did they release all the specifications of PDF1.4, although Technote 5407 documents the transparency of PDF 1.4 support.

(6) PDF 1.5

In April 2003, Adobe announced that Acrobat 6 would start shipping in late May. As always, a new version of Acrobat also brought out a new version of the PDF format, that is PDF 1.5.

PDF 1.5 introduces new functions as follows:

① Improved compaction technology, including Object Stream and JPEG 2000 compaction.

② Layers;

③ Improve support for Tagged PDF;

④ The Acrobat software itself provides more immediate benefits than the new PDF archive format.

(7) PDF 1.6 and PDF 1.7

Adobe launched Acrobat 7 in January 2005 with new PDF features, PDF 1.6 provides the following improvements:

① Improved encryption algorithms;

② Some small improvements for comments and annotation functions;

③ OpenType fonts can be embedded directly in PDF and no longer need to be embedded in the TrueType or PostScript Type 1 font form.

Adobe Acrobat 8.0 was launched in November 2006. It does not use PDF 1.7 as a preset archive format, but uses PDF 1.6 instead. For printing and prepress projects, PDF 1.3 or PDF 1.4 is sufficient. Other new functions including improved PDF/A, better organization of menus and tools, and the ability to store tables in Adobe Reader 8, and the fact that the pre-inspection engine is able to handle multiple corrections (known as Fix-ups) is another great leap. Most people approve of improving performance, especially for Intel Mac computers. An interesting development of PDF 1.7 is that in January 2008, it has become an official ISO standard (ISO 32000).

Section 5 Type Technology

I. Type of Font Library

Font library is a collection of electronic text fonts for foreign language fonts, Chinese fonts and related characters, which is widely used in computers, networks and related electronic products. TrueType and PS font libraries are commonly used in lithographic platemaking.

1. TrueType Font Library

TrueType (TT) is a new digital glyph description technology jointly proposed by Apple and Microsoft of the United States.

TT is a set of rich instructions for color digital functions to describe the outline shape of a font. These instructions include word construction, color filling, digital description functions, process condition control, retainer processor (TT processor) control, additional prompt information control, and so on.

TT uses the secondary B splines and straight lines in geology to describe the contour of the font, and the secondary B splines have first-order continuity and tangent continuity. The parabola can be precisely represented by a secondary B spline. More complex font shapes can be expressed by the mathematical characteristics of B sample length curves with several connected secondary B splines and straight lines. Files that describe TT fonts (including TT font description information, instruction sets, various tag forms, etc.) may be common to Mac and PC platforms. On the Mac platform, it is stored as a Sfnt resource; on the Windows platform, it is saved as a TTF file. For cross-platform compatibility of TT, the data format of the font file is stored in a Motorola type data structure (high in the front and low in the latter). All Intel platforms have to be preprocessed appropriately before they are implemented. The TT interpreter of Windows is already included in its GDI (Graphical Device Interface). Therefore, any output device supported by Windows can be output in TT font.

TT technology enjoys the following advantages:

(1) Truly "what you see is what you get"

TT supports almost all output devices. Therefore, all TT fonts installed in the operating system can be output at the specified resolution on the output device regardless of the system's screen, laser printer or laser imagesetter for the target output device. And most typesetting applications can lay out pages precisely according to parameters such as the resolution of the current target output device.

(2) Support for font embedding technology

Support for font embedding technology can ensure cross-system transmission of files. TT embedment technology solves the problem of file and font consistency across systems. In an application, the saved file can be embedded in a file with all the TT fonts used in the file. It enables the entire file and the fonts it uses to be easily transferred to the same system on other computers for use. The font embedding technology ensures that the computer receiving the file can maintain the original format of the file by loading the TT fonts embedded with the file, using the original font to print and modify, even if the font used by the file is not installed.

(3) Operating System Platform's Compatibility

System-level TT support is currently provided by both the Mac and Windows systems. Therefore, application files with the same name among different operating system platforms are cross-platform compatible. For example, if the PageMaker on the Mac can use all the TT fonts used in the installed file, the final output effect of the file on the Mac will be highly consistent with the output effect under the Windows.

(4) ABC Word Width Value

Each character in the TT font has its own word width value. In contrast to traditional PS, the TT interpreter for the word-width description method is more precise. TT is already included in its GDI (Graphics Device Interface). Therefore, any output device supported by Windows can be output in TT font.

In the Windows, the system commonly uses the *.TTF (TrueType) outline font library file. It can both be displayed and printed, also can support infinitely variable magnification without jaggedness in any situation. However, *.FOT is the font resource file corresponding to the*. TTF file. It is a resource pointer to the TTF font file, which indicates the specific location of the TTF file used by the system, rather than being required to be specified in the FONTS folder. *. FNT (vector font library) and *. FON (display font library) are widely used.

2. PostScript Font Library

PostScript (PS) is a page description language developed in 1985 by Adobe, which is a reorganized production based on a previous language oriented to 3D graphics. It is the interface language output by the desktop system to the phototypesetting equipment, which is designed to describe images and words. The role of PS is to record and run the image and text on the page with digital formulas, and then translate it into the required output through PostScript decoder, such as displaying on the screen or running in printer, and finally translate it into the required output through PostScript decoder, such as displaying on the screen or outputting on a printer or laser phototypesetter.

PostScript language is the most popular form of page description language in the world. It holds a large number of graphic operators that can be used in any combination, and can describe and process text, geometry, and externally imported graphics. Theoretically, any complex page can be described. Its rich graphical functions and efficient description function of complex pages have attracted the support of many publishing systems' typesetting software and graphics software. Almost all prepress output devices support the PS language, whose success has also made open electronic publishing systems widely popular internationally.

In late 1980s, the PS language also became an industry standard. PS2 was launched in 1990 and then PS3 in the same year after years of experience and feedback from many PS products. The PostScript font library technology experienced the initial Type 1, Type 3 format, and in 1990 compound font library Type 0 format (OCF) was published.

II. Front End and Back End Fonts

The front end font library refers to the display font library. The actual effect of the fonts can be shown when you use these fonts in the typesetting software; the back end font library is for the output software, which is commonly called CID font library. CID will be called when the typesetting software outputs if the software does not download fonts when generating PS files, and the quality of the font library at the backend is better than that of the font library at the front end.

The front font is a TrueType font, which is used for displaying and printing. TrueType fonts are installed in the fonts folder of the system. It is used in typesetting software. Backend fonts are the fonts used in the backend RIP, which are commonly CID fonts. Users don't need to download used fonts in the frontend typesetting software when generating results files when font library in

Chapter 2 Basic Knowledge of Lithographic Platemaking

the back end RIP covers the front end fonts.

1. TrueType Front End Display Font Library

It refers to the font library, which is installed on the typesetting host for screen display (display font library can be used on FantArt, Bookmaker, Word, etc., but non-Founder software cannot use 748-yard font library). The font library used in FantArt typesetting is the display font library, which is common for 748 yards, GB, GBK, BIG5, huge font sets, etc..

(1) Founder Lanting

It is applicable to the standard TrueType font library on the Windows 95/2000 platform and to all generic software and Founder software on the Windows platform. Three types of code, GB, BIG5 and GBK, are available.

(2) Founder Miaoshou

It can be applied to Mac TrueType font library on the Mac platform and is suitable for all generic software on the MacOS platform. It provides both GB and BIG5 coding.

Founder Lanting font library can be installed on the PC platform for typesetting design; Founder Miaoshou font library can be installed on the Mac platform for typesetting design.

2. PostScript Backend Publishing Typesetting Library

It refers to the installation of font library in the backend output device (e.g., imagesetter, printer) for typesetting, also known as PS font library (this font library is installed in the backend RIP software, such as PSPPRO, PSPNT). The typesetting library cannot be displayed on the screen. Founder's typesetting libraries can be divided into 748, GB, GBK, BIG5, and extra large font set according to their codes.

Founder' s typesetting library corresponds to Founder's display font library one by one.

(1) Founder Wenyun

It is a PostScript Type 0 font library which can be installed on the PSPNT. It can be directly installed from the PC or Mac and provides four types of encoding: 748, GB, BIG5, and GBK. Founder Wenyun font library cannot be used for output when the output device connected to PSPNT exceeds 1450 dpi.

(2) Founder Tianshu

It is the same as "Founder Wenyun" format, but can be output on devices above 1450 dpi. After the installation of "Founder Wenyun" font library or "Founder Tianshu" font library, there will be two subdirectories - Fonts and Fzdata - in the PSPNT directory.

(3) Founder CID font library

The font library, which is dedicated to the PSPNT, is available in four encoding formats:748, GB, BIG5 and GBK, as well as a set of huge font libraries. Founder CID font library is installed in the PSPNT's inhabitant directory by default.

It can be output if PSPNT is not installed the corresponding Lanting font library when "Founder Lanting" and "Founder Miaoshou" font libraries are used in the frontend typesetting software, and select the "TrueType font for Windows" option in the "Reset" dialog box. Similarly, Hanyi TrueType font library can also be installed on the PSPNT host for output. However, it should be noted that if the font is distorted in the front end typesetting software (such as QuarkXpress, FreeHand), the output of TrueType words or CID words will report syntax errors.

There will be a problem of the order of font library identification when resetting font library in the PSPNT. When the "Use TrueType font of the system option" is selected, we firstly identify the TrueType word under the system, and then identify the Founder CID and Type 1 font library specified in the "font library path". If a third party font library such as "Founder Tianshu" "Founder Wenyun" and Hanyi fonts PostScript Type 0 are installed, then the third party font library, i.e. font library under "PSPNT/Fonts", will be recognized finally. The font library, which is last identified, is used for output when these three formats font library have duplicate names.

Section 6   Basic Concepts and Methods of Imposition

It is not always available for sixteenmo, octavo and other conventional format products in work, especially for packaging boxes, small cards (certificates), etc. that are often not closed. In this case, it is necessary to pay attention to placing the finished product as much as possible in the proper paper opening range at imposition to save costs.

I. Regular Imposition

Adjustments should also be made for imposition according to the actual situation depending on the printing needs (e. g., the number), as well as the limitations of machines, such as octavo paper machines, quarto paper machines, folio paper machines, and full paper machines. Generally, octavo or quarto is sufficient, because it can be solved by overhead drying, overhead printing or front and back print on a printing machine with folio and full papers.

1. Printed in One-page

At imposition, leave a 6 mm blood in the middle (vertical centerline) stitching, i. e., 3 mm blood is left on four sides of each single page (two cuts are required). It should be noted that: If the product you make is not with a bleed picture, shading, or completely one-color shading, you can follow the method of One-page imposition, cutting once in the middle.

2. Imposition of the Envelope

Generally, when making envelopes, operators are used to grouping closures together with envelopes' folding. It is a paper-intensive approach (there is a gap that is not utilized) but with pattern continuity. Another way is to produce the cover and folding separately. In this way, paper is saved, but there is one more process-gluing. That is, we need brush the glue one more time (or one more piece of double-faced tape) when the product is finished.

3. Imposition Mode of Packaging

Generally, large packaging boxes (over octavo) are free from imposition and can be covered directly.

Chapter 2 Basic Knowledge of Lithographic Platemaking

4. Imposition for Small Packaging

Try to make the imposition as tight as possible to fully use the paper. However, there are many post-processes involved in packaging boxes, and the key is to roll the box (cut out the edges and press the creased lines). In this case, it is necessary to pay attention to that the distance between the two nearest sides should be not less than 3 mm when imposition, otherwise it will be very troublesome to make the knife mold, which even affects the product quality.

Note:

(1) Determine the size of the plate according to that of the movable parts and printing press;

(2) Pay attention to the rank-size rule and positive & negative directions of page numbers during manual imposition;

(3) Leave a blood edge, generally 3~5 mm;

(4) Accuracy can only be ensured by the alignment of each single page's number and text;

(5) Paper with different weights will be offset from folding. The amount of crawling should be properly addressed;

(6) For threefold (even for involute spline), it is not recommended for papers above 157 gram.

II. Rolling Printing Plate Edit

Rolling printing refers to printing each side of a printing plate on both sides. After printing one side, the paper will be turned around for 180° and then printed on the other side. The direction of feed edge will change when printing the second side. Two copies of the same printing can be obtained after cutting along the middle of the printing back, as shown in Fig.2-5. This method is suitable for cases where there are few printings, there are contents on the front and back sides of the print on a printing sheet, and the printing press side is relatively large. For example, for a product brochure folding,6 folding with a finished product size of 87 mm×180 mm is to be printed on the quarto machine. The size of the fully format positive paper is 787 mm×1092 mm, the size of the quarto papers is about 540 mm×390 mm, and the 6 folding is 522 mm×180 mm. In this way, two product brochures folding of 522 mm×180 mm can be placed on a sheet of quarto paper. It is possible to put the front and back contents on a quarto edition, which is printed by rolling printing. After cutting, two instructions can be obtained for one quarto.

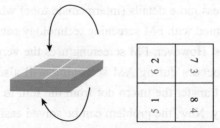

Fig.2-5 Rolling Printing Plate (Front and Back)

III. Self-flipping Layout

Self-flipping layout is a piece of printing on both sides of the paper, but the turning of the paper is common, with the same direction of feed edge, as shown in Fig.2-6. Generally, this is the way to print the cover of a sixteenmo magazine. For example, printing four covers of a magazine can make up the plate in a quarto, and then turn on the quarto machine for printing. Self-reprinting will be carried out after printing one side. Two covers are available for a quarto after the cut.

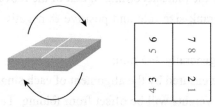

Fig.2-6 Self Flipping Plate (Side to Side)

Section 7   CTP Screening Technology

The traditional halftone screening is what we often call AM screening. It consists of a series of dots, which is regularly arranged to reflect the difference of highlight or shade by the changes of dot size. The AM screening is highly predictable and works better on the printing press than the random screening. The limitation of this technology is that the printing factory must ensure that the image does not lost dot in the brightest and darkest places. As the screen frequency increases, it becomes increasingly difficult to control the gap between the highlight dot and the dark tone mesh.

A completely different approach is used by FM screening or random screening than by AM screening. They are able to maintain the same size of the dots, but they will change the distance between the dots. Because the dot is very small and arranged very closely in many regions, the FM screening is able to express more details (intermediate tone) while reducing the opportunity of Moore streaks. Images printed with FM screening technology can often achieve the quality of continuous adjustment photos. However, FM screening uses the very small dot, so the problems of highlight dot that often occurs during AM screening will also arise in most dots of FM screening. It was difficult to transfer the micro dot from the film to the printing plate before the emergence of CTP technology. Now, the problem can be solved easily because of the emergence of new technology.

Many of the emerging screening technologies are a mixture of AM screening and FM screening. The hybrid screening uses the traditional AM screening method in most toned regions, while the FM screening technology is used in highlights and dark tone areas. Hybrid screening enables the offset printing plant to increase screen ruling in the AM screening region without placing an additional burden on printing press. Here are several of the latest screening products that are common in the market (Agfa, Esko, Fujifilm, Heidelberg, KODA and Screen).

I. Agfa

Agfa claims that its screening technology enables a combination of the advantages of AM and FM screening. Its Sublima is a patenting product that integrates both technologies into a single solution. It is also the first solution for Agfa to adopt XM patented. A total of two Agfa technologies were used during Sublima's development, i. e. Agfa Balanced Screening (ABS) and Cristal Raster FM screening.

The work method of Sublima: In the intermediate tone area, Sublima uses ABS technology to make a clear and accurate copy. In the troublesome highlight and dark tone areas, Sublima uses FM technology to reproduce the details of the image. But this software cannot be easily converted from one technology to another. It uses patenting technology to determine the switching point between the two screening technologies and can achieve a smooth transition without affecting the effect of the image. The FM screening region uses a smaller dot, but they will still be arranged according to the screening angle established by ABS. The ultimate result of this is to generate a new mesh. The screen ruling that Sublima can achieve is 210 lpi, 240 lpi, 280 lpi and 340 lpi, respectively.

All variables are taken into account by Sublima software. It can calculate the smallest dot size that can be printed on each printing press, thus allowing printing workers to adjust on the printing press-which is better than FM screening.

The internal Sublima calibration curve enables automatic compensation for different dot gains. It keeps 1% to 99% of the dot from being lost in long-page printing. The biggest advantage of this technology is still reflected in the prepress and printing. At the highest screen ruling, it can only perform RIP at a resolution of 2400 dpi. The printing effect of 340 lpi has nothing different from that of 150 lpi.

II. Esko

In September 2006, Esko launched the new Perfect Highlights flexible printing screening technology. It brings new tools to Esko users, carton processing factories and labels printing factories to let them know how to achieve the best results in the silk screen printing process and the platemaking process. The printing plant is able to set the best screening parameters for the specific printing environment, ink, printing materials and printing press, which improves the value obtained by the packaging purchaser and can also bring them a greater competitive advantage. Perfect Highlights can print 1%~2% of the highlight dots in various ways. It can bring the best intermediate coordination dark tone replication effect for users together with Esko's other screening technologies.

In addition, Esko has developed a variety of screening technologies, including SambaFlex and Groovy Screens. SambaFlex and Groovy Screens are both hybrid screening technologies. They use AM dot in most areas of the image and FM dot in the line areas of the highlights and dark tone parts to improve color saturation and the 3D effect of the image. Esko also boasts HighLine AM screening technology, which enables the production of very high screen ruling at lower output resolutions. For example, the dot of 423 lpi can be produced at a resolution of 2400 dpi.

III. Fujifilm

Fujifilm also boasts two advanced screening technologies—Co-Res AM Screening (for ordinary resolution images) and Taffeta FM Screening.

Co-Res Screening is a revolutionary, high-precision screening software developed by Fujifilm, which enables customers to print images of high screen ruling using standard output resolution. As a result, users will be able to increase the productivity of the high number of lines screening, while obtaining a more refined reproduction effect in the highlighted area.

Fujifilm also developed Taffeta, which combines itself with Image Intelligence, a perfect digital imaging technology designed by Fujifilm, to help printing factories solve problems that often arise during FM screening, and effectively increase the range and accuracy of color reproduction and reduce the generation of Moore streaks. The new Taffeta FM Screening also uses a particulate optimization algorithm. It uses optical characteristics to simulate the printing effect and uses the dot shape optimization algorithm to reduce the grain sense of the image and improve the printing adaptability of the printing sheet.

Fujifilm asserts that Taffeta is able to:
(1) Eliminate Moore streaks and rose spots;
(2) Increase the saturation of basic and inter-color;
(3) Reproduce of image details better;
(4) Improve the texture and printing applicability of the printing plate;
(5) Eliminate the non-uniformity and graininess of the printing sheet.

IV. Heidelberg

Heidelberg launched the new Prinect hybrid screening method in the UK in April 2006. This new screening method enables a combination of the advantages of AM and FM screening technology. It can bring users higher screening resolution and clearer detailed content, so as to improve the overall printing quality of the image. Prinect hybrid screening technology has been used in Suprasetter series heat-sensitive platemaking machine, Prosetter series purple laser platemaking machine and Topsetter series publishers. It can reach a maximum screening resolution of 400 lpi, depending on the CTP printing technology and the type of printing method.

Prinect hybrid screening technology can define the smallest dot for highlights and dark tone areas, and it does not allow people to use dots smaller than this size. Unlike the random

FM screening, the dot used by this technology will be arranged at an angle of AM screening. It ensures a smooth transition among the different order tuning values. In addition, Prinect hybrid screening technology distributes the mixed dot over the related angles to enhance the clarity of the detailed parts of the image and reduce the generation of Moore streaks. The Prinect hybrid screening system uses Heidelberg's own Irrational Screening (IS) technology, which not only effectively eliminates Moore streaks, but also enables skin areas that are more difficult to copy smoother. For black-and-white printing, this system can also set the black screening angle separately for a better effect.

Heidelberg also upgraded its Satin Screening FM screening technology and integrated it into the Prinect workflow to form the Prinect random screening technology.

V. KODA

The KODA Staccato software is a second-generation FM hybrid screening product, which can bring users more natural tones and smoother colors, which is very suitable in all offset printing products. The Staccato screening software can produce high-fidelity images with continuous tones. Details of these images are clear, the color gamut is wide, and the quality is high, resulting in no Moore streaks and rose spots at all. The dot size provided by the Staccato software is between 10 μm and 70 μm.

The Staccato screening software eliminates the limitations of the grayscale and the rigid connection between different tones, while also improving the stability of colors and halftone, according to KODA. Staccato also reduces image deformations caused by printing press registration errors. By using a 10000 dpi laser, KODA's Sqaurespot heat-sensitive imaging technology is able to bring users the resolution required by the Staccato screening software. As a result, it assists people to copy the tone, which is more reliable and refined in daily production. In addition to four different dot shapes for four-primary color printing, Staccate also provides six other dot shapes to support the application of KODA Spotless printing technology.

VI. Screen

Screen company released its latest hybrid screening system Spekta 2 HR. Spekta 2 enables both the exquisite printing quality of FM screening and the stable printing effect of AM screening. It can at least bring users printing quality and detail clarity comparable to the image level of 350 lpi FM screening. Spekta 2 can easily adjust the problem of dot gain, according to Screen company, because it uses 12-bit screening technology unique to screen. Spekta 2 improves people's overall control over the color and ensures the clarity and smoothing of images. Spekta 2 HR's ability to reproduce details is comparable to screen ruling with 650 lpi or more.

Section 8 Basic Knowledge of Printing Compensation and Inverse Compensation Curves

It is required for the CTP printing plate to output linearly so that the size of the dot is close to that of electronic file data to ensure the accuracy of dot transfer. However, a linear output cannot be achieved without any compensation in the case of printing plate output. The actual curve is the dot curve obtained by measuring the dot step-wedge of the printing after printing with a printing plate without printing compensation. The target curve is the dot enlargement curve of the print we prefer. It is required to compensate the printing plate accordingly when outputting it to achieve the target curve. The method of obtaining the compensation value is basically the same as the printing plate linearization.

I. Printing Compensation Curve

The purpose of CTP linearization is to establish the relationship between the dots of the digital printing file and the dots on the printing plate to control the size of the dots on the printing plate after the final exposure. When it is non-linearization, it is a straight line with a conversion curve of 45°. The process software will interpret the output according to the dot value of the digital file, thus ensuring that the dot value of the digital file is consistent with the dot value on the printing sheet. As shown in Fig.2-7, the output of the CTP device is non-linear without compensation. Therefore, it is necessary to determine the input data of the printing plate under specific exposure and development conditions so that the output of the printing plate can meet the condition of linear output.

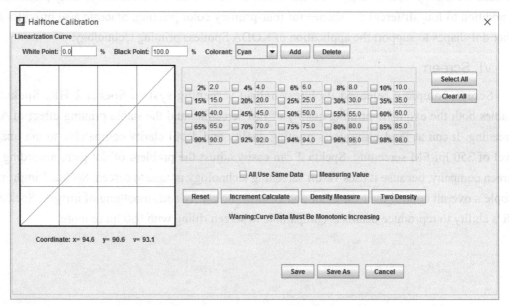

Fig.2-7   Linear Conversion Curve Without Adjustment

(1) The purpose of linearization adjustment: to make the dot size of works close to the original.

(2) Reason of linearization adjustment: There are many factors influencing the size of the dots throughout the prepress and printing process. For example, the dots will change during exposure due to the gap between the film and PS plate when plate copying; the printing press will cause dots gain due to the pressure during printing, etc.

(3) Linearization adjustment method: It must be printing plate inspection, printing matter inspection, etc.

(4) Linearization adjustment tools: SGD or DotGain of screen CTP and so on.

II. Steps to Establish a Printed Compensation Curve

1. Measurement of Printing Plates

(1) Using Illustrator graphics processing software to make a 5 mm×6 mm plot containing 30 blocks of 0%~100% color blocks. In the process software, a round dot (i.e., square round) is selected at 2540 dpi and 175 lpi without any linear curve;

(2) Set the corresponding parameters in the process: linearization curve-NONE, fine adjustment curve-NONE, printing compensation curve-NONE, printing inverse compensation curve-NONE;

(3) Screening explanation, which outputs a one-color plate on the CTP board;

(4) Dot area ratio of 9 color blocks at 50% of the plate is measured by using a printing plate analyzer to judge the uniformity of the points on the plate with an average error $\leqslant 2\%$;

(5) Measure 0%~100% of each color block to obtain the dot area ratio for the first printing plate and calculate their average.

2. Make a Linear Curve

Select "Output Device → Curve Management → PDF Screening" command in the process to enter the average value of the dot area measured by each color block on the printing sheet. Pay attention to the smoothness of the linearized curve when importing. There should be no abrupt changes, and individual control points can be ignored if necessary.

The resulting printing plate linear file is added to the "Linear Curve" and then screening outputs the printing plate measurement again to verify whether the curve meets the requirements. If the requirement is not achieved, it can be cycled several times to obtain a satisfactory printing linearization curve. The final measured value is the same as the required value, for example, at 50% dot, the measured output value of the printing plate is 50% ± 1%.

The test file to be printed will be applied to the linearized curve obtained before in the workflow is screening and the printing plate will be output.

3. Create a printing compensation curve

(1) Use Illustrator graphics processing software to make color step rulers. Dot percentage values are shown in Table 2-1 for the values in the columns corresponding to "Step-wedge Dot Percentage";

Table 2-1 Step-Wedge Dot Percentage   Unit:%

Step-wedge Dot Percentage	Expected Printing Dot Percentage (ISO standard)	Printed Plate Measurement Dot Percentage
0	0	0
2	3.1	6
4	6.3	11
6	9.4	16
10	15.7	23
20	30.5	40
30	44.3	55
40	56.9	66
50	68.4	76
60	78.3	83
70	86.5	89
80	93	94
90	97.5	98
96	99.3	100
97	99.5	100
98	99.7	100
99	99.8	100
100	100	100

(2) Determine the desired percentage value of the dot for the printing according to the ISO printing standard (shown in the second column of Table 2-1, using ISO 12647-2:2016 as the standard).

(3) Measure the percentage value of the dot on the corresponding color scale on the printing plate after output (as shown in the third column of Table 2-1).

(4) Draw the dot tone curve (as shown in Fig.2-8) and derive the dot percentage value of each step of linear compensation. The following is an example of the generation principle of the linearized compensation curve at 50% dot of the step-wedge.

① Draw a vertical line from 50% and intersect the expectation curve at point A;

② Draw a horizontal line from point A and intersect the current equipment copy curve at point B;

③ Draw a vertical line that intersects a 45° diagonal line from point B to point C;

④ Draw a horizontal line from point C, intersecting 50% of the line with point A and axis x at point D. Then point D is the correction parameter required to obtain point A on the desired value curve at 50% of the step-wedge tone. Similarly, a series of correction points can be obtained. All the correction points are connected, and the resulting curve is called a linearized compensation curve (dotted line shown in Fig.2-8).

The linearized compensation curve is selected when outputting the CTP plate to enable the output of the final printing to be consistent with the ISO standard.

Attention should be paid to the following when building and using linearization:

① There may be differences in the linearized data of the four sets of color plates when establishing linearization. This difference gradually increases with the increase of screen frequency. When we output a high number of lines printing plates, it is necessary to linearize each color plate separately;

② Screen frequency can substitute for the linearization file of four-color printing plates by using a linearized file of a single-color printing plate at 175 lpi and 200 lpi respectively. Select"Use the same data for all color plates" to generate curves to represent all color plates;

③ The purpose of the printing press compensation curve is to find the dot gain characteristic of the printing press. A compensation curve should be generated according to the target dot gain data. For example, the increase rate of 175 lpi square dot at 50% dot is generally controlled at about 15%.

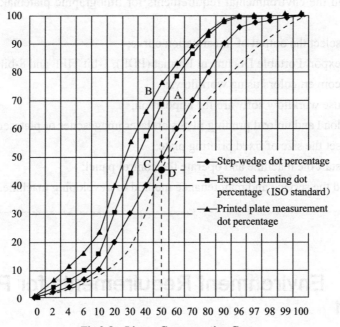

Fig.2-8　Linear Compensation Curve

Chapter 3 Printing plate output preparations

 Objectives:

1. Understand the environmental requirements for lithographic platemaking production operations;
2. Option to select the output of linearization curve;
3. Ability to export Portable Document Format (PDF), 1-bit TIFF and 8-bit TIFF;
4. Ability to convert colors using workflows;
5. Ability to use workflow software for imposition;
6. Ability to load and unload printing plates on laser imagesetter or platesetter machines;
7. Ability to set the size of fixed printing plate;
8. Ability to start the machine and operate the control panel;
9. Understand the service and maintenance method of developing machine.

Section 1 Environment Requirements for Production Operation

The environment of lithographic platemaking production operation consists of two parts: the process output environment and the CTP equipment output environment.

I. Process Output Environment

1. Server (Take CRON Yinyihuitong for Example)

The CRON Yinyihuitong server program is installed on a server. Visible to users is a program called "Process Server", whose function is to start and shut down the CRON Yinyihuitong Server.

A database system is installed on the server. The Yinyihuitong Server program uses this database to store and manage all project information. The database system is generally started automatically with the start-up of the computer system.

Working data is stored on the server. All original page files and permanent or temporary data files generated during the project are stored on the Process Server.

2. Process Workstation

The Yinyihuitong client program is installed on several process workstations. Users perform their daily production via the process client software, such as creating projects, collecting source files, designing hand-folding plans, digital proofing, printing for output, etc.

Folding layout design program is built in the client program. The users cannot directly launch the folding layout design program, but can only start the program by creating a new layout or editing and modifying the layout in the process client.

There is also a free imposition program built into the client program. Users cannot directly start the free imposition program. They can only start the free imposition program by creating a free imposition template in the process client and submitting the loose page files to be imposition to it.

II. CTP Equipment Output Environment

(1) Location requirement of the CTP equipment workshop: It is best not to choose places with large equipment on the upper or lower floors or nearby. The CTP equipment may be influenced by noise or shock due to large equipment operation, as shown in Fig.3-1.

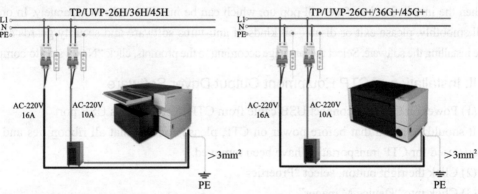

Note: The grounding of the equipment showld comply with the installation guidelines, and the ground resistance should be measured by a professional measuring in strument to control it $\leqslant 0.5\ \Omega$. All power connections are made by twisting. Users need to lay the power cord along the route shown in the diagram. Wires not noted are copper cored.

Fig.3-1 Wiring Diagram of CTP Platemaking Machine

(2) The air indicators in the CTP working room are required to meet the National Quality Class II standard, i.e., the API value is greater than 50 and no more than 100.

(3) The API value refers to the air pollution index, when it reaches 50, it is the primary standard and when it reaches 100, it is the secondary standard.

(4) For work sites with constant temperature and humidity, the optimal temperature is required at 23℃ $\pm$ 2℃ and the optimal humidity is required at 50% $\pm$ 10%.

(5) The level of the workplace is required to be below 4 mm.

(6) The requirement of the power supply is that the unidirectional line cannot be less than 4 mm^2, and it cannot be less than 25A for the unidirectional line air switch.

(7) The ground resistance requirement is that a 4 mm^2 diameter ground wire is connected to a galvanized metal bar with a diameter of not less than 10 mm. The metal rods should be buried 1.5 m or more under the wet soil, and the resistance of the equipment to the ground is less than 0.5 Ω.

(8) The water requirement for the printing equipment is to be sufficient enough and clean. If it is insufficient, a booster pump is required; for areas with poor water quality, a filter device should be installed.

Section 2   Installation of Output Process Software and Device Driver Software

I. CTP Output Process Software's Installation (The New Laboo 5.X Output Software as an Example)

The users first need put the software installation CD into the computer optical CD-Rom drive, and then the installation interface will pop up, which can be installed and run separately. In order to install smoothly, please exit or disable all kinds of anti-virus software and security guards software before installing the software. Select the language according to the prompts, click "Next" step to complete.

II. Installation of CTP Equipment Output Driver Software

(1) Power on CTP and connect USB cable from CTP to computer's USB port

It should be noted that before power on CTP, please confirm that all ribbon ties and fixed supports used for CTP transportation have been removed.

(2) Click the right button, select "Proerties".

(3) Click into "Device Manager".

(4) Click on the "Action" from the top menu, select the "scan for hardware changes" for new hardware detection. The setup process is shown as Fig.3-2.

(5) After seeking the newly displayed uninstalled USB driver after connection, right click it and select "Update Driver Software", as shown in Fig.3-3.

(6) Select "Browse my computer for driver software" to manually install USB driver, and do not select 'Search automatically for updated driver software', as shown in Fig.3-4.

(7) Manually select the USB driver location, as shown in Fig.3-5, to install the USB driver of Windows 7 64bits. Please find the Laboo\ USBDrv128B\USBDrv128B\sign Drv\win7_amd64. Click "Next" after confirming that the USB driver is correct.

Chapter 3 Printing plate output preparations

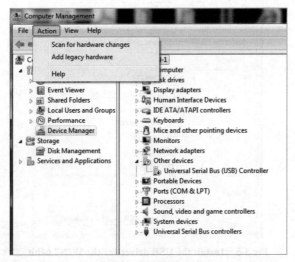

Fig.3-2   Setup Process

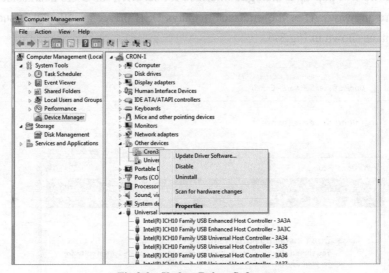

Fig.3-3   Update Driver Software

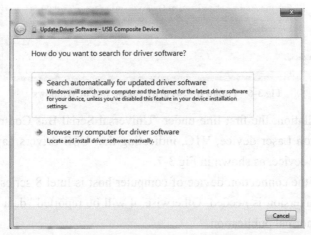

Fig.3-4   Browse through the computer to find the driver software

| Computer to Plate of Lithographic |

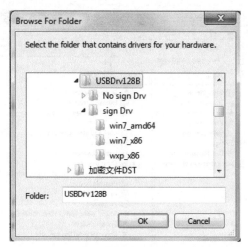

Fig.3-5　Install the USB driver for the WIN7 64bit

(8) Windows will pop-up a dialogue interface for security, the users need select "Install this driver software anyway". After the interface displays USB driver installation is successful, click "Close" to exit, as shown in Fig.3-6.

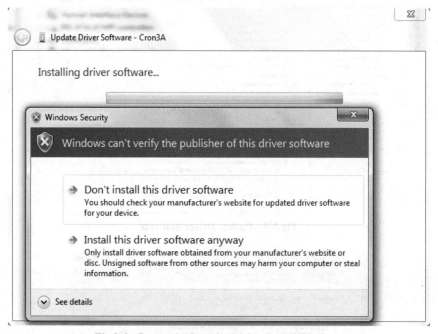

Fig.3-6　Popup the Security Dialog Box of Windows

(9) After installation, the first line under "Universal Serial Bus Controller" in the Divice Manager Shows: Cron Laser device, V1C, indicating that USB drivers have been succesfully installed for the CTP device, as shown in Fig.3-7.

Special note: if the connection device of computer host is Intel 8 series motherboards, then PCI interface for conversion is needed. Otherwise, it will be reported "data transmission error", affecting the device normal publication.

Chapter 3　Printing plate output preparations

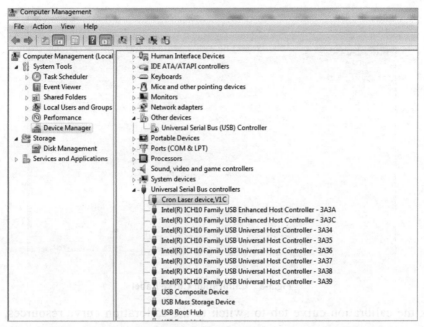

Figure 3-7　The "Universal Serial Bus Controller" of the Device Manager

Section 3　Linearization of Printing Press Output

If you switch from traditional platemaking process to the CTP direct platemaking process, you will find that CTP prints are darker than traditional prints (under the same printing conditions). This is because in the traditional platemaking process, the data file is first output to the film through the imagesetter, and the dots are transferred from the film to the positive-type PS plate. Due to factors such as the oblique rays of light and the inaccuracy of the dot edges during printing, the dot area on the printing plate is smaller than the dot area on the film, and then the dots will be enlarged after printing; However, there is no film in the CTP platemaking process, and the dot area on the printing plate is the same as the dot area of the data file. Therefore, through the linear calibration of the CTP platemaking machine, the print quality is consistent with or better than that of traditional printing.

I. Create a Calibration Curve

The calibration curve tab of the Resource Library is used for the calibration curve used in the creation process, as shown in Fig.3-8.

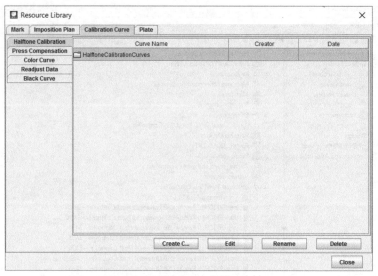

Fig.3-8　Calibration Curve Label

Click the calibration curve tab to switch to the calibration curve resources, including dot linearization, printing compensation, color linearization, secondary data, and single black curves. Dot linearization and printing compensation are used in the output templates, and color linearization, secondary data and single black curves are used in the proofing templates.

In the calibration curves window, you can toggle the interface of each curve by clicking the tab on the left. On each interface, you can create, edit, rename, and delete curves.

Click "New Curve" to pop up the "Dot Linearization" dialog box, as shown in Fig.3-9.

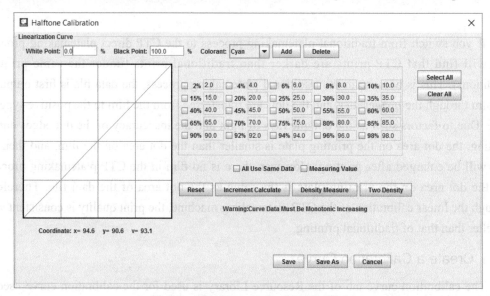

Fig.3-9　"Dot Linearization" Dialog Box

The default values for "White Point" and "Black Point" are 0 and 100, respectively. In general, users are not suggested to change the number unless it is in a special application. If the White Point are filled with w% and the Black Point are filled with k%, then the dot will be output for w% for places with 0% blackness (blank) on the page; the dot will be output for k% for places with 100% blackness (field), with the middle part adjusted according to the linear curve.

The "Color Plates" by default are cyan, magenta, yellow and black. A spot color plate can be added or removed through clicking the "Add" or "Delete" button. After clicking "Add", the "Add Spot Color" dialog box will pop up, which can be added according to the name of the spot color plate in the file.

Linear curves can be set for each color plate individually. On the left is a visual graph of the linearized curve, and on the right is the parameter for setting the linearized curve, which can be set by 2%~98%; The set value takes effelt after Clicking the checkbox in front of the set value. Values are usually filled in according to the measured results.

The method of measurement: Prepare the documents for testing, which generally are the grayscale step-wedges of different levels from 2%~98% shown in "Dot Linearization". It is exported without the Dot Linearization curve set in the output templates. The actual percentage value of the dot for each color block on the grayscale is measured and recorded by using a transmission/reflection densitometer. Fill in each set value in turn with the results measured just now.

(1) Same data is used for all color plates: It means that each color plate uses the same data after being selected.

(2) Data are measured values: The measured values of the linearized curves entered are adjusted to standard values.

(3) Reset: Restore all values to their original state.

(4) Incremental calculation: Incremental calculation of linearized curves. Enter the measured value based on the current curve to make incremental adjustments to the curve. After clicking, the "Incremental Calculation" dialog box will pop up, as shown in Fig.3-10.

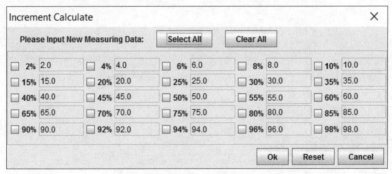

Fig.3-10 Incremental Calculation

(5) The value of the linearized curve must be monotonic in increment: It indicates that the value entered by the linearized curve must be monotonic in increment, otherwise an error will occur.

(6) Save: Click "Save" after the linearization curve is set. Enter a name for the linearized curve in its dialog box, then click "OK". The new curve will appear in the "Dot Linearization" tab, which can view the creator and creation time of each curve.

(7) Edit: Double-click the curve to open the "Edit Curve" dialog box and set it. You can

click "Save As" to save it as a new curve without affecting the previous one.

(8) Rename: Reset the name of the curve. After selecting the curve, click "Rename", the "Rename" dialog box will pop up, and then set the name.

(9) Delete: Delete the selected curve.

II. Implementing Linearization in Process Software

In the "Calibration Curve" setting tab, both Dot Linearization and Printing Compensation can be set. Both curves can be created according to different requirements if they are to be in the repository.

1. Dot Linearization

The linearization curve is used to correct the errors caused by equipment and environment factors at the output. Typically, the same output device is at different resolutions so the linearization curve will vary when outputting with different network wires. It is best to build a linearized curve for each resolution.

After "Dot Linearization" is selected, click "Browse" to pop up the "CURVES Selection" dialog box, as shown in Fig.3-11.

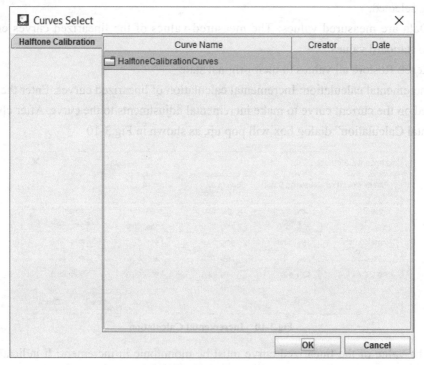

Fig.3-11 Dot Linearization

In the dialog box, the Dot Linearization curve resource created by users in the resource library is displayed. Select the required curve and click "OK".

2. Printing Compensation

Printing compensation is used to compensate for the dot gain caused by printing pressure. The printing compensation curve is set in the same way as "Dot Linearization".

Chapter 3   Printing plate output preparations

Section 4   Operation of Output Process Software

I. Normative Documents

1. Submit the Source File to the Normalized Processor

The process of assigning a task in a project to a processor node is called "submit". When a project is submitted to a node template, it means that the relevant page data and parameters in the template are handed over to the corresponding Job Ticket Processor (JTP) on the server for processing. The results of the processing will be displayed to the client, and new data generated by the processing will be stored in the folder of the corresponding project on the server.

The first step after the successful addition of the file is to normalize the file, i. e., to normalize the source file into a standard single PDF file.

Select the file that is ready to be added, click the "Submit" button, or right-click the file and choose the "Submit" option to open the "Submit" dialog box. Templates that can be submitted by the source file are listed in the dialog box. Select the normalization template to be submitted and click "OK" to normalize, as shown in Fig.3-12.

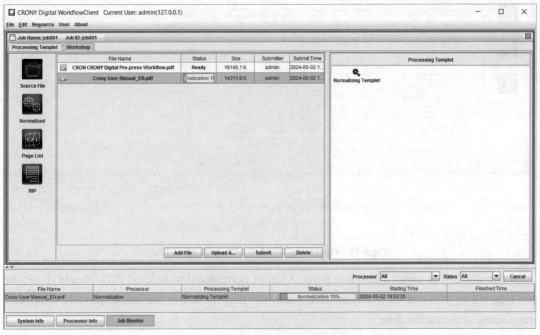

Fig.3-12   Normalized Interface

After the successful submission, the file will be normalized. The status bar in the window enables users to view the progress of the file processing, as well as the detailed processing progress of the file in the project monitoring. All operations submitted in the process can also be submitted directly by using computer dragging. The steps are as follows: First, select the file to be processed, and then hold down the left button and drag it directly to the corresponding template on the right. Now the pointer turns to a plus sign and the file will be automatically processed accordingly.

The "Cancel" operation will terminate the operation being processed by the current file. Normalization of the file can be canceled in the source file. Select the file that is being normalized and right-click the "Cancel" option to stop the normalized file and display it as "Termination" in the status bar. All canceled projects can be performed in "Project Monitoring". Select a processing task that is being performed and click "Cancel".

2. Specification File Window

In the "Specification Files" window, the source file is normalized, a standard single PDF file is ganerated after. Here, the file can be previewed and can be submitted to the free imposition templates and output templates. Click the "Specification File" icon, the interface will be changed to the operation window of the standard file, as shown in Fig.3-13.

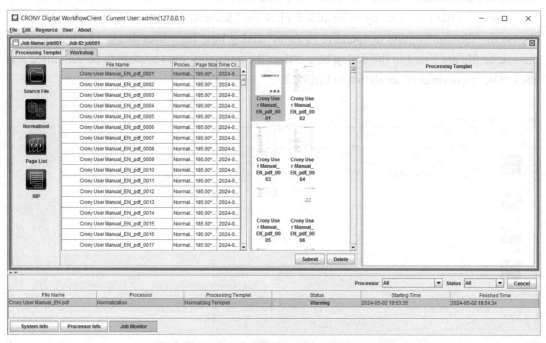

Fig.3-13 Operation Window of Specification File

The window is divided into two parts. A list of normalized individual PDF files is shown on the left, where you can view information about the file, including file name, process templates, page sizes, and creation time.

(1) File name: Name of a single PDF file for the generated standard. The file name is named after the name, the number of pages, and the type of the source file. The page number of each file and the source file type are clearly shown by the name.

(2) Process template: We can find out where the file was generated by the name of the templated used for normalizing the source file.

(3) Page size: The page size of the file is shown in the form of width × height, in mm units.

(4) Created time: The time when the file was generated.

A thumnail of the file is shown on the right, with a corresponding thumbnail for each file. The file is selected on the left and the thumbnail on the right is also selected, by which we can preview the general content of the file.

In the "Specification Files" window, preview, submit, cancel, delete, etc. can be performed on the file.

3. Preview Window

Users can sketch preview through a file thumbnail or get a detailed preview by opening the preview dialog box.

Select a file and right-click it to select "Preview" or double-click the file in the dialog box that is displayed; the "Preview" dialog box will pop up, as shown in Fig.3-14.

Fig.3-14 Preview

In the dialog box, operations such as Zoom In, Zoom Out, Ruler, Drag, Pipe, Print, Mirror, and Reverse can be performed on the image.

There is the preview toolbox at the top of the dialog box, where different tools can be selected for preview; it is the preview area of the file in the middle, which shows the preview image of the file; the operation information of preview is displayed in the status box below, including the scaling multiple, the length and angle when using the scale, and the value displayed by the pipe.

The buttons in the preview toolbar are arranged in the order from left to right, with functions as follows:

(1) Zoom In: After clicking the "Enlarge" 🔍 button, the file is magnified exponentially, with a maximum of 16:1. In the status bar of the dialog box, you can view the zoom level of the file. Press and hold the "Ctrl" button when the cursor changes to a palm during the preview, and the cursor will change to a magnifying glass. Now, you can click and enlarge directly, or select a region in the window to zoom in on the image by pulling a box to local preview (this operation cannot be used when using the "ruler").

(2) Zoom Out: After clicking the button 🔍 , the file will be reduced by multiple to a minimum of 1:16. Similar to zooming in, press and hold the Shift key when the cursor change to a palm, and the cursor will change to a reduced state. Now you can click the "Zoom Out" for preview directly. In any state, the "+", and "−" keys on the keyboard can be pressed directly to zoom in/out the image preview.

(3) Ruler: After clicking the button 📏 , the cursor will change to a cross-shape, where the corresponding object can be measured. Select the starting point of the measurement object, left-click and hold it; move your cursor to the end of the object to be measured. The specific values of the length and angle of the object will be displayed in the status bar of the dialog box. Press and hold the Shift button and you can make straight measurements horizontally and vertically.

(4) Drag: Click the ✥ button to change the cursor to the state of dragging when the cursor changes to a palm. Drag to preview the magnified image (preview dialog box, which is opened as the default when dragged).

(5) Continuous splice: Preview the images by continuous stitching of files. This action is just a preview for continuous output. After clicking the button, the image will be automatically generated into three images that are vertically spliced together, where users can see whether the results of the images are correct after they are stitched together.

(6) Rotate List Box: In the Rotation Drop-down List Box, select the appropriate degree for preview to view the effect of the file after rotation.

(7) Mirror Check Box: When the check box is selected, the file can be mirror previewed.

(8) Reverse Check Box: After the check box is selected, the file can be previewed forward and backward and reversed. The "Forward" and "Backward" options only work for previewing multiple files. After clicking, you can view a preview of the previous or subsequent file. "Forward" and "Backward" operations can be controlled by using the left and right direction keys on the keyboard.

(9) File Display Box: At the end of the toolbox, the name of the preview file is displayed. For multiple files, you can view and select the files of the preview in the drop-down list.

4. File Information Window

The File Information window enables you to view file information. Right-click and select "File Information" or press and hold "Ctrl+I", then the File Information dialog box will popup, as shown in Fig.3-15. In the "File Information" dialog box, the file name, page size, and color separation information are displayed.

Chapter 3 Printing plate output preparations

Fig.3-15 Viewing File Information

Display the color separation information of the file, including parameters such as global color separation, color name, extension error, spot color to CMYK, etc.

(1) Global Color Separation: All color plates in the file are displayed in the corresponding colors, including four colors of C, M, Y, K and spot color plates.

(2) Color Name: Display the names of all color plates.

(3) Extension Error: Adjustment of page size when exporting files to the output according to the needs of different devices. Expand or shorten the page to the left and right according to the set value.

(4) Spot Color to CMYK: It works primarily for documents with spot color. For spot color, when selected, the spot color in the file is converted to four color formats of CMYK for explanatory output. This operation should be set conjunction with the "Color Separation" parameter in the output templates.

(5) Update All Pages: It indicates that the parameters set in the dialog box will affect all the specification files. After the setting is finished, click "OK" to complete the setting of the file information.

Press and hold the Ctrl and Shift keys while selecting multiple files. Press and hold Ctrl+A to select all files; click "Delete" or right-click and select "Delete", the system will pop up the corresponding prompt according to the selected files; click "OK" to delete the file. The file that is in use when deleting it cannot be deleted. Files cannot be restored after deletion.

5. Submit Normalization Documents

In the Normalization window, imposition and output can be performed on the normalized standard PDF file. This is the second step in the submission process. Click the "Submit" button or right-click the "Submit" option to open the "Submit" dialog box (see Fig.3-16), in which all the templates that can be submitted are displayed.

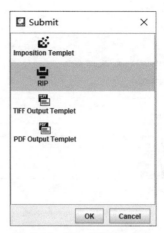

Fig.3-16 Submit Dialog Box

(1) Submit to free imposition template

In workflow software, free imposition can be performed on the submitted file. After setting, perform the "Imposition" (as shown in Fig.3-17), and the software will automatically generate the large plate document into the "Printing Style" interface of the process.

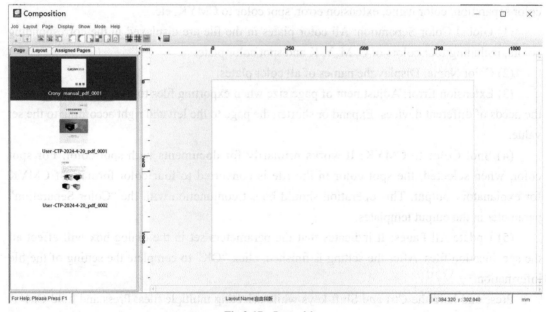

Fig.3-17 Imposition

(2) Submit to the Output Template

The output operation in the process includes PDF output, TIFF output, digital printing output, and CTP output. PDF output and TIFF output interpret a file directly into a network-shared folder; digital printing output and CTP output have their own outputs machine, and then the process interprets the file to the output machine.

The path set in the template must be connected properly and have file creation rights when submitted to the PDF output and TIFF output templates. In this way, the output will be proper.

Chapter 3　Printing plate output preparations

It is necessary to first determine whether the output machine has been successfully started when submitting to the digital printing output and CTP output templates. The icons on the right will be shown in grey when they fail to boot. After submitting, the system will be prompted "the follower failed to be turned on and the output cannot be performed"; After successful start-up, the icons will be shown as dark blue so that the output action can be performed. Click "OK" after selecting the corresponding template for submission; After submitting, the progress of explaining the output can be found in "Project Monitoring".

After the file export is finished, it will be automatically generated into the corresponding shared folder or in the corresponding exporter.

Screening parameter setting is completed by the output template of the production output process, in which the CRON JionUS process software is achieved by the RIP setting of the CTP output template (as shown in Fig.3-18).

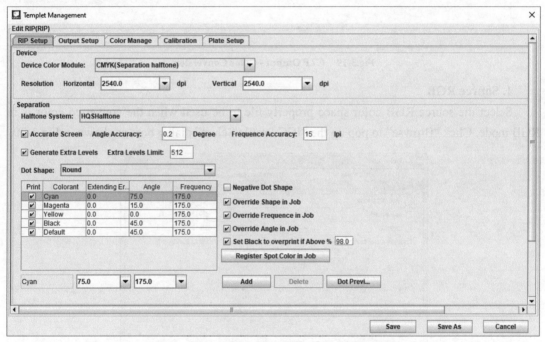

Figure 3-18　RIP Setup of CTP Output Template

Dot shape corresponding to the "Dot Shape" can be clicked in the drop-down list in dialog box. The dot angle and screen ruling are modified in the list according to different color plates. For spot color, the "Add" button enables you to set the screening angle and number of lines.

II. Workflow Conversion Color

It has its own color management module built into the process. The ICC Profile is used to achieve the proper output of different color mode files on different devices through the color characteristic description files of each device. There are three drop-down list options on the "Color Conversion" setting page (as shown in Fig.3-19); each option is available for picking a color property description file.

Computer to Plate of Lithographic

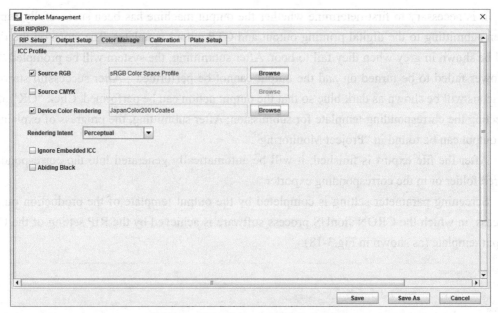

Fig.3-19 CTP Output - Color Conversion

1. Source RGB

Select the source RGB color space property file to be used when the converted color is in RGB mode. Click "Browse" to pop up the "ICC Profile Select" dialog box, as shown in Fig.3-20.

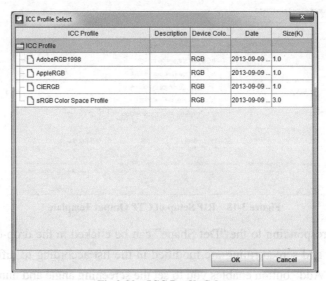

Fig.3-20 ICC Profile Select

In the dialog box, the ICC property file resources with RGB color space in the resource library are listed. Select the required ICC file and click "OK" to finish. The process defaults to "sRGB Color Space Profile".

2. Source CMYK

Source CMYK is the source color space property file that is used when the converted color selected is in CMYK mode. The method of selecting a file is the same as the Source RGB item.

Chapter 3 Printing plate output preparations

3. Device Color Space

Select the destination device color space characteristic description file to be converted to. Usually, a property file for the printing press should be selected. Select file method as above.

Color management is used to perform the color conversion between two devices. In case the value of a color on device A is Ca and you want to convert it to the value Cb on device B, the color characteristic description files of both devices A and B should be used. The color characteristic description of device A is called the "Source Color Space", and the color characteristic description of device B is called the "Target Color Space". In other words, to correct the color of a printing press to be in line with that of a traditional prototype machine, the prototype machine is the "Source Device" and the printing press is the "Destination".

4. Intention to Color Conversion

The desired color conversion intention can be selected according to different output requirements, which is Perceptive by default.

Perceptual: Perceptual mapping is performed to the color space of the device from the color space independent of the equipment; all colors are compressed in the same proportion, and the colors outside the color field are mapped within the color gamut range of the equipment; this way is suitable for color conversion of images.

Absolute: Absolute Colorimetric will keep those in the color field constant and compress those out of the color field to the boundaries of the target color gamut. This method is suitable for the accurate reproduction of color.

Relative: Relative Colorimetric is used for color accuracy and media-related reproduction Mapping white and black fields to the target color gamut can change the brightness.

Saturation: The saturation method increases the purity and saturation of colors but suffers from poor color reproduction. This approach is suitable for focusing only on the reproduction of brightly colored graphics.

[Ignore the embedded ICC]

This option indicates that embeddedness in the file is ignored during the interpretation of the output. This item is usually selected to avoid problems that cause inaccurate colors or words turn to four colors of black due to the improper embedding of the ICC file in the source file generation.

[Keep it solid black] [Maintain solid black]

This option is designed to avoid overprinting difficulties caused by the conversion of black graphics or text objects to four-color black during conversion from RGB to CMYK.

III. Font Library Management and Setting

The font library is a key resource in the process. The correctness of font library installation affects the correctness of the file after normalization. The process installs the standard font library automatically during installation. For PS files, users do not need to install fonts if they automatically download fonts at the time of generation. However, PDF files sometimes do not have embedded fonts, so they should be installed manually for the file to be properly interpreted.

After opening the "Resource Library", click the "Font Library" tab; all the fonts that have been installed are shown in the window as a list.

(1) Installation font: Click "Install" and the "Installation Font" dialog box will pop up; there are four basic options at the top of the window.

Source installation font library path: Set the location where the font library file is to be installed. The function of the "Installation Font Library" is only used to load the TrueType font library. Click the "Open" button to browse folders to help users find the font library file path.

TrueType is installed: This option describes the format in which the TrueType format font library is loaded into the process. Options such as "Local JoinUS Fonts by default" (default option), "PostScript CID 2 font by default", "PostScript TYPE 42 font by default", etc. are available.

The process reorganizes the TrueType font library into a new font library format when it installs the font library. In general, CID 2 font library can be selected if the Chinese font, Japanese font, and Korean font are to be installed; the TYPE 42 font library can be selected if the English font is to be installed. It is not required to select CID coding. The font resources and installation of fonts as shown in Fig.3-21 and Fig.3-22.

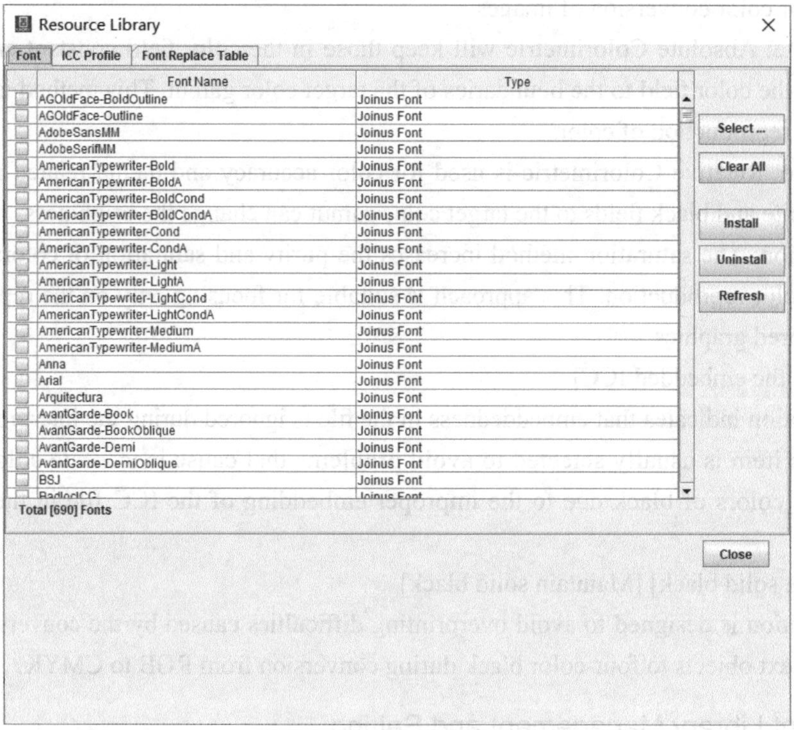

Fig.3-21 Font Resources

A unique set of font library formats, known as "JoinUS Fonts", is constructed in order to properly handle the problems of word filling, vertical typing, etc. The most common TrueType font library can be installed in this format. Therefore, the "Local JoinUS Fonts by default" is the default option. Font library can be installed properly by selecting this option in general.

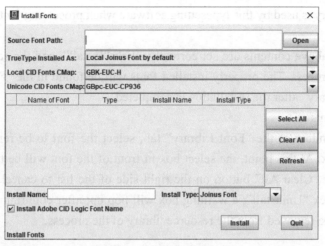

Fig.3-22   Installation of Fonts

Local CID Coding Table: This option sets the mapping method from local encoding to the CID font library. Local encoding refers to the regional encoding of each region. For Chinese, there are GB2312, GBK, BIG5, etc.; Japan and Korea also have their local codes. There are many options for different regions, but users in the Chinese mainland will generally choose the default option to install the font library (whether it is GB2312 or GBK font library). The default is GBK_EUC_H.

Unicode CID Code: TrueType font library is generally organized font library according to Unicode, so the mapping method from Unicode to local encoding needs to be set. The users can select the default selection. The default is GBpc_EUC_CP936.

Font List Window: Its middle is the font list. When the font library location is specified from the "Source Font Library Installation Path", the system will start to search for this folder, and the searched fonts will be listed in order. It also shows the number of font files searched at the bottom, and lists various information about these installed fonts, including "font name", such as HYf0gj, "type", or "installation name", which is named after installation into the process, and "installation type", which is the type selected in "Install the TrueType as".

The corresponding information of the font library will appear in both the "Installation Name" and "Installation Type" columns at the bottom of the font list window when a font in the font list window is clicking. Users can change the installation name and type in these two bars. However, changes are generally not required for users.

There is a selection box in front of each font listed in the font list window, click the check box to determine whether the font is installed. The "Select All" and "Clear All" buttons on the right side of the list are used to assist in selecting the fonts to be installed. The selected font selection box will be marked with a " √ ". "Refresh" button is used to update the contents of the font list window.

There is another item at the bottom of the window.

Registered Adobe CID logic font library: It indicates that the logic font library is generated at the same time as the font library installation. This function is set for some typesetting software.

A logical font library is used by this typesetting software when processing fonts, which is selected by default.

After all the above contents are set correctly, click "Install", and the selected font will be installed into the process. The properly installed fonts will be found in the "Font Library" tab of the "Resource Library" after the installation is completed.

(2) Unload fonts

In the list window of the "Font Library" tab, select the font to be removed through the previous check box. At this point, the select box in front of the font will be typed " √ ", or click the "Select All" or "Clear All" button on the right side of the list to cancel the unloaded font. After selecting, click "Uninstall"; a warning box will pop up; after confirming, click "OK". The selected font will be removed from the resource library of the process.

IV. Imposition

JoinUS Hezhe is the hand-folding imposition plate design software in the JoinUS digital workflow, which is responsible for creating a folding plate for hand-folding imposition in the process, including paper size, plate, small page attributes, template marking, folding method, etc.

1. New Hand-Folding Layout

Right-click in the page list to open the "New Layout" dialog box, as shown in Fig.3-23.

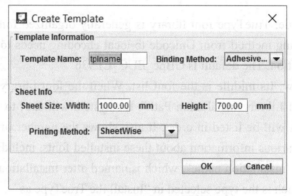

Fig.3-23   New Hand-Folding Layout

(1) "Template Information" setting box

Template Name: Name of the plate file.

Bookbinding method: Saddle Stitching, Perfect Binding, and Free Binding

Perfect Binding: It can be for projects such as paperbacks, and all sections are overlaid in parallel when bookbinding.

Saddle Stitching: It can be for projects such as brochures, outlines, and catalogs. It is a bookbinding method that nests the sections together, as shown in Fig.3-24.

Free Binding: It is for non-folding templates, such as posters, stitching, etc. In the Free Binding, templates with different page sizes and directions can be combined on the printing for collage projects.

Chapter 3　Printing plate output preparations

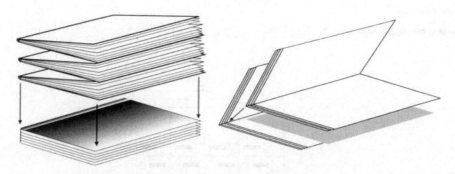

Fig.3-24　Saddle Stitching Bookbinding

(2) "Printing Sheet Information" setting box

Paper Size: Size of the actual printing paper.

Printing methods: single-side, double-side, self-turning and roll.

Single side: corresponds to single side printing, printing is only on the front, which is often used for posters, business cards and labels.

Double side: The most common way of printing on double side is to print the front and back of the paper with different plates. On the front and back sides of the plate, the front and back positions of the small plate are mirrored and balanced, as shown in Fig.3-25.

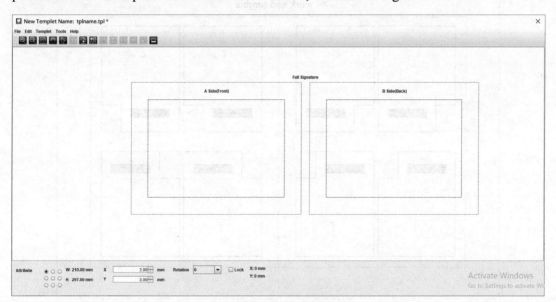

Fig.3-25　Double Side

Self-turning: for auto turning printing mode, both sides of the imposition are located on the same printing plate. The front and back sides of imposition are symmetrical left and right corresponding to the vertical centerline of the paper, as shown in Fig.3-26.

Rolling: for rolling printing mode, both sides of the imposition are located on the same printing board. The front and back sides of the imposition are symmetrical up and down corresponding to the horizontal centerline of the paper, as shown in Fig.3-27.

Computer to Plate of Lithographic

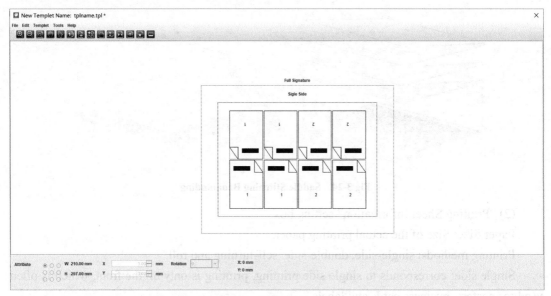

Fig.3-26　Self-turning Plate

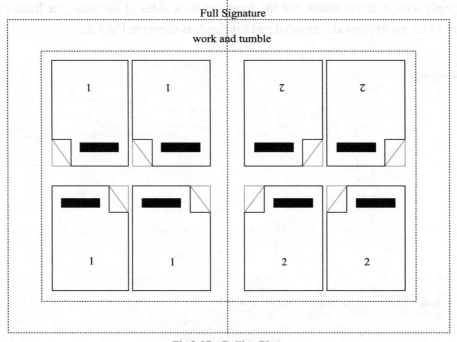

Fig.3-27　Rolling Plate

After setting the template and printing sheet information, click "OK" to open the imposition design interface as follows.

The Hand-Folding interface (as shown in Fig.3-28) consists of four parts: Menu column, Tool panel, Design panel and Attribute panel. The black dotted frame in the design panel is the paper size we set.

Chapter 3　Printing plate output preparations

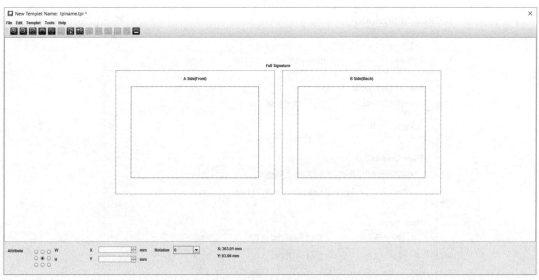

Fig.3-28　Hand-Folding Interface

2. Toolbar

The toolbar contains 14 buttons, as shown in Fig.3-29. Sort from left to right in the figure, the functions of each button are shown as follows.

Fig.3-29　Toolbar Buttons

(1) Zoom In ![] : Enlarge the imposition view;
(2) Zoom Out ![] : Narrow out the imposition view;
(3) Fit to the window ![] : Zooms the box view to fit the window;
(4) Drag ![] : Move the window view;
(5) Select ![] : Used to select pages or marks;
(6) Set page number ![] : Used to automatically add page numbers in sequence.
(7) Add an individual page ![] : Used to add individual pages to the plate.
(8) Add imposition ![] : Used to add hand-folding imposition to the plate.
(9) Page spacing ![] : Control to display and hide the page spacing display box;
(10) Top-bottom centering ![] : Center the imposition top and bottom on the paper.
(11) Left-right centering ![] : Center the imposition left and right on the paper.
(12) Display/Hide Pages ![] : Control to display and hide pages.
(13) Display/Hide marks ![] : Controls the display and hiding of marks;
(14) Feed edge location ![] : Control the feed edge position of the paper.

3. Create an Imposition

Execute "Template" → "Create Imposition" to open the Create Imposition dialog box, as shown in Fig.3-30.

Computer to Plate of Lithographic

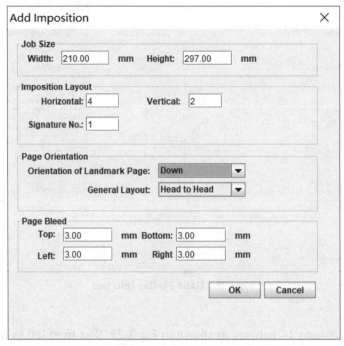

Fig.3-30 Creation of Imposition

(1) "Finished Product Size" setting box

Enter the width and height of the size of the finished product. The imposition pages on the printing sheet are of the same size. The size of a page cannot be changed, but the size of all pages in imposition can be changed.

(2) "Imposition Layout" setting box

Enter the number of horizontal and vertical imposition pages to be added to the printing side.

Number of staples: It is used to control the imposition layout of a printing sheet with multiple staples.

(3) "Page Direction" setting box

Single-page direction: Specifies the way the imposition page is oriented (relative to other pages), where it refers to the direction of the page at the top left. The remaining imposition pages in the group are oriented relatively to the upper left pages.

Overall Style: It is used to set the direction of the overall page, including head-to-head, toe-to-toe, head-to-toe.

4. Set the Page Number

After imposition is established, the page number of all pages will default to be 1. Users need to set the page number according to the folding method, which can be in two ways.

(1) Set the page number manually

Double-click the page to be modified to pop up the "Set Page Number" dialog box. Users can also set the page number on the front and back of the lower sheet.

(2) Set page number automatically

Lower sheet numbers are automatically added with the "Set Page Number" tool.

Chapter 3 Printing plate output preparations

Select the "Set Page Number" tool and click the lower sheet to set the page number, which will automatically set the page number for the front and back of the lower sheet. Page numbers increase with clicking in sequence. Left-click the 🔲, and the "Set Page Number" dialog box pops up to set the starting page number.

5. Set Page Spacing and Margins

For the imposition created, users can set the page spacing and margins of the imposition. On the tools panel, click the "Display/Hide Page Spacing" icon. The lower sheet spacing values appear on the top and left of the imposition.

The value of page spacing can be modified by left-clicking the value of page spacing to open the "Set Page Spacing" dialog box (Fig.3-31). After modifying the page spacing, you can center the imposition up and down and left and right on the paper by left-clicking the "Top and Bottom Center" and "Left and Right Center" icons in the tool panel.

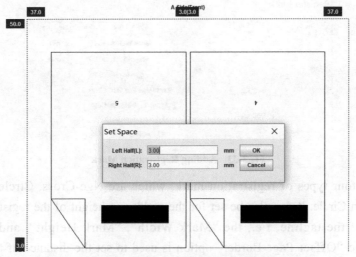

Fig.3-31 Set Page Spacing Dialog Box

6. Add Marks

Add the Cut, Registration, Folding, Side Gauge, Scalp, Text, Repeat and Custom marks for imposition, and set the options as shown in Fig.3-32.

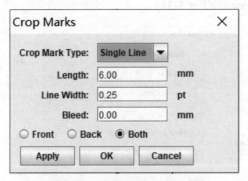

Fig.3-32 Adding a Cut Mark

(1) Cut mark

The "Cutting Type": It enables you to choose both single and dual-line types for the cut token.

The "Length": The marked length of the cut, 6 mm by default.

The "Line width": The marked width of the cut, 0.25 pt (0.88 mm) by default.

Offset Bleeding Value: The distance between the cut line and the blood site that typically is set to 0, which means that the cut line is added beyond the ventilation.

The "Front Side", "Back Side" and "Front-Back Sides" indicate on which side of the printing sheet the tag will be displayed.

(2) Registration mark

The registration mark is shown on the left side of the dialog box, as shown in Fig.3-33. The function is to add a registration mark to the imposition page. It is used to check whether overprinting of the printing press is accurate.

Fig.3-33   Adding Registration Mark

There are four types of registration mark, which are Nge-Cross, Circle Mark, Hollow Circle, and Solid Circle. It can also be set for the width and height of the registration mark and for the width of the tagline, i.e., the "Mark Width", "Mark Height", and "Line Width" commands. The "Offset Page Border" option is used to set the distance of the tag from the page, with a default value of 3. To ensure that the registration mark is in addition to vomiting, the offset value here is generally the blood value of imposition. The "Front Side", "Back Side", and "Front-Back Sides" options are used to indicate on which side of the sheet the tag will be displayed. The registration mark is usually added on the front and back sides.

(3) Fold marks (as shown in Fig.3-34)

Fig.3-34   Add Folding Mark

Its parameters are as follows.

"Length": The length of the folding mark;

"Line Width": The width of the folding mark line;

"Center Cross": Located at the center of the folding of hand-folding imposition, it is used to check whether folding is accurate. Here is a choice of whether to add a crosshairs mark;

"Mark Color": It is used to set the color value of folding marks;

"All Color Separations": The tag will appear on all color separation plates, and the color value of the color can be set in the subsequent text box;

"Printing Color": The marks only appear on the four-color plate of CMYK, and users can set the color values for marking CMYK color separation respectively.

V. Export to File

1. TIFF Output Template

TIFF output template is a file that is interpreted and output as a TIFF format, with the corresponding dialog box shown in Fig.3-35. The template is divided into five parameter setting tab pages, which are RIP setting, output setting, color conversion, curve correction and printing plate setting. All four except for the output setting are in line with the parameter setting in the CTP output templates.

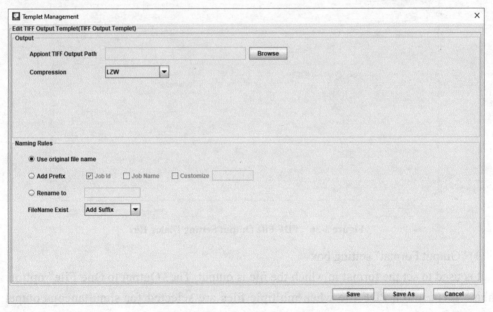

Fig.3-35　Edit TIFF Output Template

The output setting differs from the CTP output template by setting the TIFF file output path and file naming rules in the output setting tab.

(1) "Output" setting box

Sets the output path of the TIFF file. Click "Browse" button and a path selection box will pop up. Select a local path or network share path.

The local path can be selected when the server and the client are installed on the same machine. A shared path is required when it is not on the same machine.

(2) "Naming Rule" setting box

There are three naming rules: use the original file name, add a suffix, and rename the file.

"Use Original File Name": The output TIFF file name is named directly according to the file name after the normalized or imposition in the process.

"Add Suffix": Suffix of job ticket number, project name, and custom name can be added after the file name, respectively. The suffix is _ 1 after the same file name in order.

"File Rename": Set a custom name for the exported TIFF file.

"Processing File with the Same Name": For the file of the same name after output, suffix and overwrite can be added to the file.

2. PDF File Output

Output the specification file in the process or the imposition large plate document as a standard PDF file. After the file is exported successfully, a folder with the same name as job ticket number will be created at the specified path to store these PDF files. Fig.3-36 is the "PDF File Export Setting" dialog box.

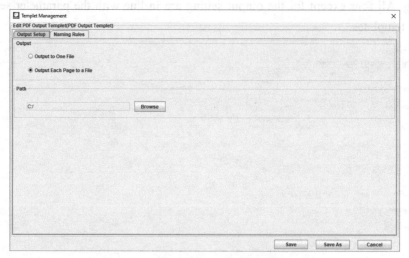

Figure 3-36   PDF File Output Setting Dialog Box

(1) "Output Format" setting box

It is used to set the format in which the file is output. The "Output to One File" option is to combine them into a PDF file when multiple files are selected for simultaneous output; the "Output Each Page to a File" option is to output multiple files as a single file when multiple files are selected for simultaneous output.

(2) "Path" setting box

It is used to select the storage path of the PDF file.

Click the "Browse" button and select a network-shared folder. The local path can be selected when the server and the client are installed on the same machine. The options in the "Single Page File" setting box are the same as the TIFF output setting.

Section 5 Platemaking Machine and External Equipment

I. Basic Construction of Platemaking Machine

CRON's CTP platemaking machine thermal series TP-46, TP-36 and TP-26 series are exampled as follows: CTP platemaking machine consists of optical part, machinery part and electrical part, as shown in Fig.3-37.

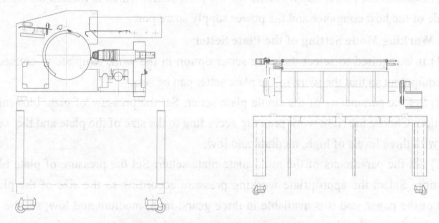

Fig.3-37 Structural of CTP Machine

Optical part: include laser, fiber coupling, close-arranged head, optical lens, optical energy measurement, etc.

Machinery part: machine frame, light drum, wallboard, plate sending department, head opening/closing department, plate load roller department, plate unload department, typesetting department, screw guide track department, scanning platform department, etc.

Electrical part: Encoder, main/auxiliary servo motor, stepping motors in actuators, vacuum pump, master board, terminal board, laser drive plate, sensors, etc.

Connect CTP platemaking machine to the computer through a USB cable which is located at the USB wiring port on the lower right side of the casing and is fixed with data interface bolts to ensure a reliable connection.

II. Plate Setter

CTP plate setter is a device that realizes paper printing separation and automatic plate feeding and is used to realize the automatic board supply operation of CTP platemaking equipment. Its structure is shown in Fig.3-38.

1. Connection of Plate Setter

(1) Connect the communication line for the plate setter, which is located on the port side of the USB wiring of the host computer;

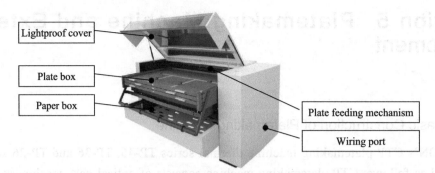

Fig.3-38 Structure of the Plate Setter

(2) Connect the power supply cable for the plate setter, which is located on the USB wiring port side of the host computer and the power supply spare port.

2. Working Mode Setting of the Plate Setter

(1) It is required to select the plate setter option in the work template to connect the plate setter equipment so that the work of the plate setter can be set;

(2) Set the parameters of the single plate setter: Set the pressure of plate blowing and plate absorption. Choose the right work pressure according to the size of the plate and the weight of the paper, with three levels of high, medium and low;

(3) Set the parameters of the multi-plate plate setter: Set the pressure of plate blowing and absorption. Select the appropriate working pressure according to the size of the plate and the weight of the paper, and it is available in three gears: high, medium and low; set the parameters for the plate box of the plate setter and it supports a maximum of 5 printing boxes.

III. Built-in Perforation

Built-in perforation is an optional peripheral, which is a module that can be integrated and installed inside the CTP equipment to complete synchronous overprinting and punching. The structure of the built-in perforation is shown in Fig.3-39.

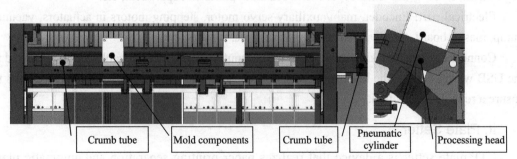

Fig.3-39 Built-in Perforating Machine Structure

1. Connection of Built-in Perforation

(1) Click the "Auxiliary Setting" in the software operation interface to bring up the "External Device Setting" dialog box and select the "Punch a Bridge" setting interface.

(2) Select "Built-in Perforation" in the dialog box and check it in the check box of "Select

Bridge".

(3) There is no need to set the bridge type and left-side bridge parameters.

2. Built-in Perforation Work Setting

(1) For equipment containing built-in perforation function, it is required to check to "Punch Bridge" option in the work template, which can be set for built-in perforation;

(2) Equipment No.: Select "Built-in Perforation";

(3) Inlet direction and outlet direction: Built-in perforation is not required to be set;

(4) Mold selection: It is used to select the group number of the current work mold, and only single group work is supported. A maximum of 5 groups of molds are supported.

3. Built-in Perforation Maintenance

(1) Observe regularly whether the aluminum chips are burrs. Modify the mold if necessary to extend the service life;

(2) Check the lubrication and sealing of the cylinder rods regularly;

(3) Replace the cylinder seal every 24 months.

IV. Punch Bridge

Punch Bridge is an optional peripheral equipment, which is a machine that can automatically complete overprinting and punching according to the setting of CTP equipment. Its functions are to automatically match the specification plate used by CTP platemaking equipment, automatic overprinting printing, four-way plate input/output, and automatically connect to the printing equipment. The structure of BGP Perforation for the Gap Bridge is shown in Fig.3-40.

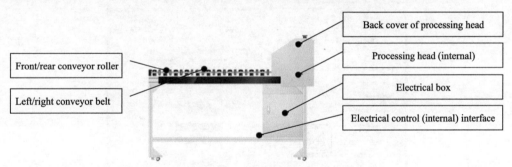

Fig.3-40 Structure of BGP Punch Bridge

1. Punch Bridge Auxiliary Setting

(1) Click "Auxiliary Setting" in the software operation interface to call up the "External Device Setting" dialog box and select "Punch Bridge" setting interface (see Fig.3-41).

(2) Select a valid bridge number according to the number of BGP equipment connections (check in the check box of "Select Bridge"), and up to 4 groups of bridge connections are supported;

(3) Set the correct "Bridge Type" according to the model.

Computer to Plate of Lithographic

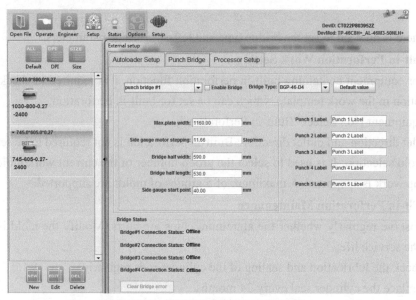

Fig.3-41   Auxiliary Setting of Punch Bridge

2. Work Setting of Punch Bridge

(1) The punch bridge option should be checked in the work template for equipment connected to BGP Punch Bridge, which can be set for punch bridge, as shown in Fig.3-42;

Fig.3-42   Working Setting of Punch Bridge

(2) Equipment No.: It is used to select the order number of the current BGP equipment. The equipment supports the online use of multiple groups of BGP equipment;

(3) Inlet direction: Set the direction of the plate entering the BGP equipment. The equipment is available in four directions;

(4) Outlet direction: Set the direction of the plate leaving the BGP equipment. The equipment is available in four directions;

(5) Mold selection: It is used to select the group number of the current working mold of

Chapter 3　Printing plate output preparations

BGP equipment, and only single group work is supported. This equipment supports a maximum of 3 sets of molds for installation; the short bridge is lifted up and the back-end position is aligned with the processor inlet.

V. Dust Collector

1. Product Structure

The vacuum cleaner is used to remove the burnt dust produced during the exposure process and reduce the pollution of the burnt dust to the machine. At the same time, the burning smell is diluted to reduce the smell of the space.

The structure of the vacuum cleaner mainly includes three areas: air pump working area, electrical control area and filter element installation area, as shown in Fig.3-43 (above). The filter element installation area contains one layer of initial filtration, two layers of odor removal filtration, and two layers of superfine filtration, through which five layers of filtration it can reduce the burnt dust and taste produced effectively by the exposure of the equipment, as shown in Fig.3-43 (below).

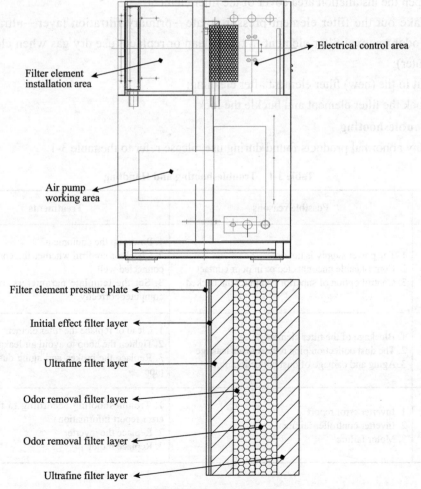

Fig.3-43　Dust Collector Structure

2. Use and Maintenance

(1) Regular cleaning: Replace the primary filtration layer every month, clean the vacuum dust and odor removal filter element every three months, and replace the vacuum filter element / odor removal filter element every six months.

① If the device is used in a dusty environment or the plate is dusty, the maintenance and replacement of the filter element cycle should be shortened.

② If the filter element is improperly maintained, it may cause the decline of dust and odor removal effect, which will seriously affect the dust removal effect of CTP plate output, and even do the damage to CTP laser system.

(2) For the high operating temperature of equipment: please maintain the environmental temperature of the equipment to avoid the containment of the ventilation and cooling area, which will affect the operation and even cause equipment damage.

(3) Ensure that the dusting pipeline is unobstructed, check whether the pipeline is aging / damaged / distorted every six months, and replace it in time when necessary.

3. Replacement method of the filter element

① Open the installation area cover of the filter element;

② Take out the filter element pressure plate—primary filtration layer—ultrafine filter element—odor removal filter element in turn, clean or replace (use dry gas when cleaning, do not use water);

③ Put in the (new) filter element after cleaning;

④ Lock the filter element and buckle the lock.

4. Troubleshooting

For any abnormal products found during use, please refer to the table 3-1.

Table 3-1 Troubleshooting and Handling

Faults	Possible reasons	Treatments
Machine out of work	1. The power supply is not on 2. Control cable unconnected or in poor contact 3. Control option of superior computer unchecked	1. Power up the equipment 2. Check or confirm whether the control wire is connected well 3. Set the template parameters of the upper computer correctly
Decreased attraction	1. Blockage of the filter element 2. The dust collection pipe relates to air leakage 3. Aging and damage of dust collection pipe	1. Clean or replace the filter element 2. Tighten the hoop to avoid air leakage 3. Replace the broken and aging dust collection pipe
Fault indicator on	1. Inverter error report 2. Inverter controller failure 3. Motor failure	1. Troubleshooting according to the inverter error report information 2. Replace the inverter 3. Replace motor

Chapter 3 Printing plate output preparations

Section 6 Processor

Development is required through the processor after the CTP printing plate is outputted, and then the CTP printing plate imaged by development can be used for printing. Therefore, the state of the processor needs to be adjusted and confirmed before the printing plate is outputted. For best performance, the processor and its surroundings must stay clean. The processor should be cleaned at the beginning and end of each working day.

Observe the following rules for cleaning:

(1) It is forbidden to use hard brushes, grinding materials, solvents, acidic or ordinary alkali solutions to clean the roll shafts and other components;

(2) Only the specified chemicals can be applied to clean the processor;

(3) It is not allowed to use sandpaper, detergent fabrics, or other grinding materials to remove dirt or substances from processor components.

(4) Clean the processor with a white lint-free cloth.

I. Inspection of Equipment

All components of the processor should be carefully inspected at least once a week. It should be noted that the main power supply needs to be turned off before opening the processor cap. Table 3-2 shows what is to be viewed when checking the processor and performing related maintenance tasks.

Table 3-2 Contents of Inspection Processor and Execution of Maintenance Tasks

Failure parts	Phenomenon of failure	Treatment method
Development Part	There is serious crystallization residue in the developing tank	Drain the developing tank and rinse with developing cleaner
	The developer has a low liquid level	Add replenishment
	Circulation pump does not circulate	Check whether the spray pipe is blocked and clean it; Check whether the circulating pump is working properly
Wash part	The spray pipe is blocked	Clean the spray pipe
	Less water	Increase the inlet switch valve, or increase the pump
	The water is not clean	Increase the water purifier
Gumming part	The gluing hose is blocked	Clean the gluing hose
	The glue has more impurities	Replace the glue
	The glue has a higher concentration	Add water to increase the dilution ratio
Drying part	The plate is not completely dry after it comes out of the processor	Increase drying temperature
Roller part	Abnormal operation	If the roller in the processor is abnormal, you can adjust the tightness or add Vaseline; If the external roller of the processor is abnormal, you can add mechanical butter
Condense part	Poor cooling of the developer	Check the condensate level and add the condensate; If the condensate has many impurities, replace the condensate

II. Development Part Maintenance

It should be noted that the developer is corrosive. Wear protective goggles, nitrile gloves and protective clothing and dispose of discarded development solutions according to local regulations. First, exhaust developing liquid tank. Second, clean the development part. The development part should be cleaned when performing maintenance and service. Before cleaning the developing tank, it must be ensured that the developing liquid is run out and that the drain valve is turned on. The drain pipe must be empty when put into the drain channel or empty container. Finally, replace developing filter (as shown in Fig.3-44). It should be noted that protective clothes, glasses and gloves are needed when performing part development.

1-Developing filter Outlet Valve (open); 2-Developing filter Inlet Valve (open); 3-Press—release Button

Fig.3-44　Developing Filter

III. Maintenance of the Washing Section

The maintenauce methods of the washing section are as follows.

First, drain washing tank. Second, clean the Washing section tank. The wash section tank should be cleaned when performing periodic maintenance and service. Finally, replace the washing section tank's filter.

IV. Maintenance of Gumming Section

The maintenance of gumming section: perform the following at the end of each working day to prevent the sizing from hardening:

① Place the rest of the gum back into the container;

② Wash and rinse the gumming section.

V. Condenser Section

The condenser is located in front of the processor chassis and is used to cool the developing liquid in the development section.The condenser consists of the following components (as shown in Fig.3-45): refrigeration unit (compressor, heat sink and fan), condenser related components.

> **Chapter 3   Printing plate output preparations**

1-Compressor; 2-Heat sinks; 3-Fan; 4-Pump; 5-Drainpipe; 6-Drain valve

Fig.3-45   Condenser

Chapter 4 Output of Printing Plate

Objectives:

1. Output parameters can be set according to the process sheet;
2. Be able to check the digital files processed by the Raster Image Processor (RIP);
3. Be able to complete the development of the printing plate;
4. Master the solutions to problems that arise during the printing plate output.

Section 1 CTP Printing Plate Output

The steps to start the CTP platemaker are as follows: ① It is necessary to turn on the UPS power supply, and observe whether the pointer of the temperature/humidity meter is in the green position; ② Turn on the red power switch on the right side of the CTP; ③ Turn on the computer power supply connected with platemaking machine; ④ Turn on the touch switch on CTP plate-making machine and wait for 5 minutes for the machine self-test; ⑤ Turn on the driver software and check the status of the software and machine (as shown in Fig.4-1).

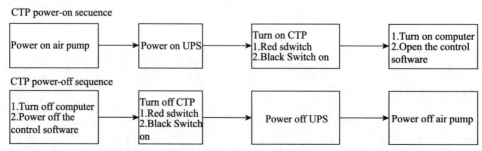

Fig.4-1 Start-up & Shut-down Process of the CTP Platemaker

Chapter 4 Output of Printing Plate

I. Start-up

When using the CTP platemaking machine, turn on the UPS power supply and turn on the air conditioner. Then, turn on the air compressor and watch whether the pointer in the green position; turn on the power supply to the equipment using the cam switch, and is turn the cam switch to the position shown in Fig.4-2.

Fig.4-2 Cam Switch

Open the circuit breaker up to the ON position for equipment using a mini circuit breaker; connect the equipment power supply (3-phase power supply circuit breaker on the right side), as shown in Fig.4-3.

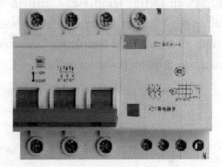

Fig.4-3 Power Supply Disconnection

Turn on the touch switch on the CTP and wait 5 minutes for the machine to self-test. Ensure that the platemaking machine's device driver is installed properly in the "Computer Lines" → "Device Manager" deployment directory. During proper installation, the following hardware "Cron Laser Device, V1C" can be found under the universal serial bus controller (as shown in Fig.4-4).

The software will not be able to connect to the device if it is not installed correctly. At this time, the updated driver can receive artificial assist in the installation of the platemaking machine device driver. The software driver files are stored in the software root-"USBDrv 128B" folder. Please update manually until the device driver is properly installed.

After the software is properly installed, double-click the Laboo 5.1.0 software shortcut icon on the desktop to pop up the start-up interface. The Laboo output is started after the start-up program is loaded.

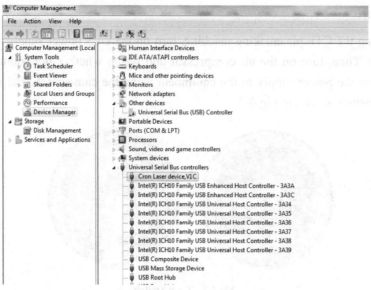

Fig.4-4　Confirmation for Platemaking Machine Device Driver Installation

II. Set the Initial Template

The initial template needs to be set for new softwares. Click the "New Template" dialog box to set an initial template: Confirm input of parameters such as "Template Name", "Resolution", "Carrier Material", "Plate width", "Platemaking Energy", "Exposure Speed", etc. according to the allowed template of the model specification setting (as shown in Fig.4-5). The parameter description is shown in Table 4-1.

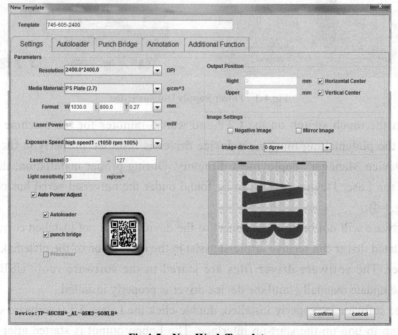

Fig.4-5　New Work Template

Chapter 4 Output of Printing Plate

Table 4-1 Set Parameter

Name	Description
Template Name	Set the template name; suggest specification naming: Plate Width—Resolution—Plate Thickness. For example: 510-400-2400-0.135
Resolution	Set the image resolution used by the template (different models are configured to support different resolutions)
Carrier material	Set the material of the template to use, supported by PS plate, and film (different configuration of the model, supporting material is also different)
	Note: Error in the material setting will cause problems such as off-focus and equipment jitter, which will affect the imaging quality
Plate width	Set the plate width size of the template; refer to the schematic diagram at the bottom right for the dimension direction; the thickness shall be subject to the measurement (The maximum and minimum plate width are determined by the model)
Laser power	The laser power is a manual set value, which depends on the energy requirement of the plate material and whose value does not exceed the ex-factory power of the equipment
	Note: Manual set is not required if the check is automatically adjusted according to the energy of platemaking, it is recommended to choose
Exposure speed	The speed of the light drum during exposure can be selected according to the actual production setting. Its maximum speed is determined by the platemaking energy, maximum power and maximum speed limit
Number of laser paths	Set number of paths 0 ~ 128 (number of laser paths determined by the model)
Platemaking Energy	Platemaking energy is determined by the characteristics of the plate and is provided or measured by the plate material supplier
Plate setter	The automatic plate set option, If checked the template works including automatic plate supply by the plate setter (model decision)
Punch Bridge	Punch option, If checked the template works with an automatic punch (model determination)
Processor	Processor options (determined by the model)

New work template: Click "New Template" button to bring up the "New Template" dialog box, and set a working template according to actual work needs; confirm after setting and create a hot folder to facilitate subsequent batch operation.

III. Edit Work Template and Add Project

Edit Work Template: After the work template is set, re-edit the template can be realized through the edit template function if it is necessary to change template parameters. Template name, carrier material, and resolution parameters will be locked and unable to be edited during template editing, which is limited to changes of adjustable parameters.

Add tasks by adding TIFF file: Add TIFF file by clicking the "Import File" button, select *.TIFF and *.PPG file projects in the corresponding directory, and select the file to be used and the corresponding template below (to ensure that the parameters are consistent); after confirmation, the file will appear in the output list of the software (as shown in Fig.4-6).

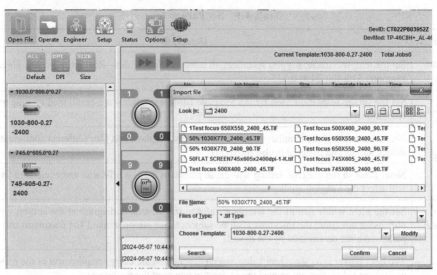

Fig.4-6 Add File

IV. Inspection before Output

Check the file carefully; select the correct file to output the CTP plate material according to the order; and perform the last regular inspection on the TIFF file (angular line, color code, size, color separation, leading edge, stitch).

Set the plate material output parameters in the process software; set the publishing power, rotation speed, image position and other parameters according to the actual demand of the output plate material.

V. CTP Loading

According to the software operation instructions, select the correct template, perform "Template Adjustment", and adjust the equipment template to the corresponding position; check the flatness of the plate material for black/white dots; check the size of the plate material and prompts from the computer to operate; align the lower edge of the plate material with the edge of the machine during the loading, and then pressed the press button; it is required to take it gently and lay it levelly on the processor when unloading.

The steps of setting down the plate material as follows:

① Set down the plate material in the corresponding position of a simple plate set, with the right side of the plate close to the middle position and leave about 1 mm by the specification;

② Place a batch of printing materials (maximum of 50 copies) on the set platform of the equipment, with the front-end close to the front regulation, the right side close to the middle position by the gauge, and leave by about 1 mm.

VI. Plate Output

1. Set Output Parameters

Parameter setting for the system is shown in 4-7.

Chapter 4   Output of Printing Plate

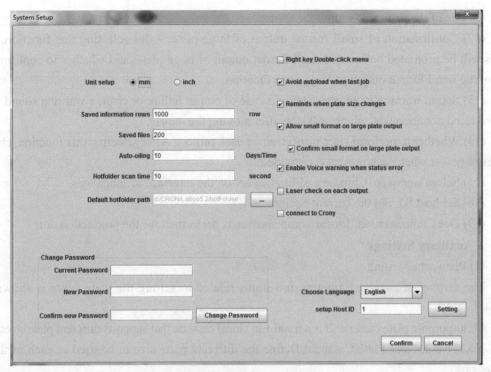

Fig.4-7   Parameter Setting

(1) For unit setting, i. e., the calculation unit of the output image, mm or inch can be selected;

(2) Number of reserved information rows: Any feedback information will be automatically retained when the equipment executes the action. This function can set the number of rows to be retained. Right-click on the feedback Information interface to remove all reserved information. 1000 rows of information saved is shown here.

(3) Number of files reserved: 200 of files saved is shown here.

(4) Start time of the slow rotation of the light drum: Start the slow rotation of the light drum when the equipment is in standby state.

(5) Slow rotating light drum stop time: The time when the light drum stop rotating when it is in a slow rotating state.

(6) Hot file scan interval: The system automatically searches for hot folder periods, with a default value of 10 second.

(7) Default path of hot folder: It is possible to set which folder is specified by the default state of hot files.

(8) Double-click in the "Information Bar" to pop up the shortcut menu: Select this option and then right-click in the "Information Bar" dialog box to choose action.

(9) Plate set is not automatic for the last project: plate setter is not automatic for plate set after the last project is finished.

(10) Frame switching is reminded when automatic plate set output: the software will be prompted when different frames switch.

(11) Allow the large plate to output a small format: Select this function to use a plate larger

than the size of the pattern to output the format.

(12) Confirmation of small format output of large plate: After selecting this function, the users will be prompted before the small format output of large plate, and whether to confirm the use of the small format output of large plate function.

(13) Sound warning in an error state: In case of output failure or error, a warning sound will be produced continuously, which will stop after moving the mouse.

(14) Whether to perform laser inspection on each project: After selecting this function, check the laser power when each project is outputted.

(15) Select language: Chinese and English are on the interface of languages.

(16) Set host ID: Set the current equipment ID.

(17) Dot Compensation: Linear compensation is performed for the production dot.

2. Auxiliary Settings

(1) Plate setter setting

The display of usage parameters and status related to setting the plate setter is shown in Fig.4-8.

Multi-purpose plate cassette: It is a multi-functional cassette that supports different plate sizes;

No.1, No.2, No.3 Plate Cassette: Define the different plate size to be used in each of these cassettes;

paper blowing settings: strong, medium, weak, according to different types of slip paper;

paper sucking settings: strong, medium, weak, according to types of slip paper with different absorption pressure of the paper.

Supplier status: Online status of the current supplier.

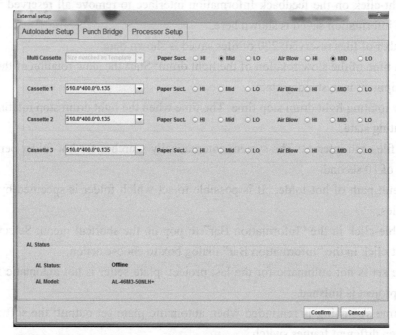

Fig.4-8 Display of Using Parameters and Status related to Plate Setter Setting

Chapter 4 Output of Printing Plate

(2) Punch Bridge (see Fig.4-9)

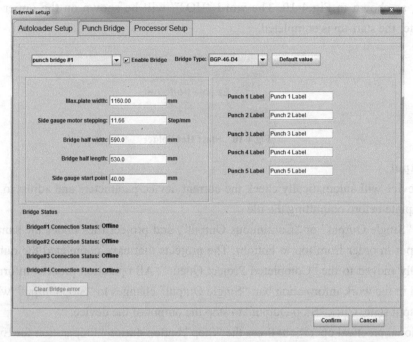

Fig.4-9 The Display of Using Parameter and Status Related to Punch Bridge

① Select gap bridge: Select gap bridge according to project needs;

② Side gauge maximum print width: The largest print width for current gap bridge work;

③ Side gauge motor step: The number of impulses moving by 1 mm of the side gauge step motor;

④ Half width of gap bridge: The current gap bridge supports half of the maximum plate width;

⑤ Half-length of gap bridge: The current gap bridge supports half of the maximum plate length;

⑥ Start position of side gauge: The start position of side gauge pull plate in the working state;

⑦ Names of the mold 1, 2, 3, 4, 5: User-defined name of the current gap bridge mold;

⑧ Gap bridge status: Current gap bridge online state.

3. Add Files

Two ways are available to add files for the Laboo follower: One is from the CRONY process; while the other is to add a TIFF file directly. The steps of adaing TIFF are as follows:

(1) Select "Open" in "Toolbar";

(2) Select the *.TIFF file and select the template to be applied to the buckling in the pop-up dialog box; After the successful addition, the information area of the software interface will display "Succeeded in Adding File".

(3) All added projects appear in the "Wait Output Queues" of the file.

Note: You can add files to the output via a hot folder if the output template is set up and activated.

Activation method of hot folder: Select a template and right-click to select "Enable hotfolder", as shown in Fig.4-10. The word "HOT" will be shown on the upper left of the template after the start-up is completed.

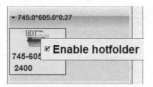

Fig.4-10　Start Hot Folder

4. Output

The device will automatically check the current device parameters and adjust to the current project template before outputting the file.

Click "Single Output" or "Continuous Output", and project files under the same template will be output in order from top to bottom. The projects that have completed the output will be automatically moved to the "Completed Project Queue". All operation process information will be reflected in the work information bar. "Single Output" changes to "Stop Output" when the file is in the output state. Click "Stop Output" to stop the output of the device.

Additional practical operation: Press the "Stop" button during "Preparation for Exposure" or at the initial stage of exposure processing if you want to cancel the production after the previous edition. After pressing the stop button, the device will not be automatically removed. You need to click the command in "User Operations" to remove the printing plate from the device.

For errors such as light shift, just press the "OK" button, and follow the prompts for some errors.

5. Precautions

(1) Copy the TIFF document to the corresponding size publishing folder; confirm whether the machine display size corresponds before printing the plate.

(2) Ensure that your hands are free from dust and water before taking the plate from the box.

(3) The printing plate and the lining paper should be taken out together during the printing process, otherwise, the layout is prone to friction and white lines are generated.

(4) Don't force your fingers too hard during taking out and dropping off the plate to prevent folding the print. Be careful the layout should not be hit by external objects.

(5) Before placing the plate on the platform of the loading, the plate should be checked for levelness, special attention should be paid to whether the edges of the plate are buckled to ensure levelness, otherwise the platemaking machine stuck plate will be easily caused.

(6) Attention is required for proper position dimensions when setting down the plate; otherwise, it will be stuck during plate loading.

(7) The cover should be gently buckled Avoid too much force, otherwise, the laser will vibrate easily and a long blue line will be left on the plate.

(8) Load the waiting plate should be completed before the 100% exposure of the previous one.

(9) Please adjust the sizes of the corresponding output mode before pressing the "Plate Load" button when connecting printing plates of different sizes, otherwise the printing plate will be stuck and the adjustment sensor parts will be distorted, followed by an error, and its door will fail to be opened or closed.

(10) Before topping a file, it is necessary to confirm whether the waiting plate loaded in the machine corresponds to the size of the file. In short, ensure that the waiting plate corresponds to the size of the next document to be published.

(11) Avoid being inclined when pushing it into the processor, otherwise, it will be stuck during the printing process. The "Emergency Stop" should be pressed immediatly if it happens.

(12) Push a little hard to ensure that the front end of the cone fully grasp the printing plate, otherwise, the printing plate will stagnate on the platform ahead.

(13) Collect the plate in time to avoid scratching its surface by overlaying two plates.

(14) A lining paper should be linked between two plates to prevent wear and scratch when collecting the plates.

(15) Both the original and the finished plates should not be exposed to light for a long time; after the plates are out, put the box cover back on.

6. Daily Work and Maintenance

(1) Registration is required for output.

(2) Check the bills of the previous day every morning, and the supplement plate is also required to be checked.

(3) Press a test plate every morning and night. The deviation range should be controlled within ±1%. For example, the large deviation of the dot may be due to the early weakening of the drug property because of the recent excessive publishing, or the lack of replacement of the developing liquid for too long, or the decrease of the developing liquid effectiveness due to the water flowing into developing tank resulting the blockage of the external circulation drain of the tank. Take the following measures: Change the newly developing liquid. If the external circulation drain of the sink is blocked, components such as brushes should be removed and then the drain should be unchoked. In addition to the cause of the developing liquid and machine, the error may be that the document itself is malfunctioning.

(4) Ensure that the ground is clean.

(5) The indoor temperature should be controlled properly at 21 ~ 25 ℃.

VII. Turn off CTP

Turn off the CTP printing equipment with/via the device driver from the computer; Turn off the CTP platemaking machine, the computer, and turn off the power supply.

The use of platemaker is noted as follows:

(1) CTP platemaker cannot be turned off during operation, otherwise, there will be a plate stuck;

(2) Place the CTP plate with the film facing up and in parallel to the printing port, instead of being inclined or backward.

Section 2 Set the Resolution and Screening Parameters of the Press

I. Setting Press Output Resolution in the Direct-to-plate Process

The setting of the output resolution of the printing plate in the direct-to-plate process is accomplished by setting the parameters of the output process template, for example, the setting of the CRON process software is achieved by normalizing the template parameter setting.

The normalized template is to interpret the source file to a standard single PDF file. The template contains four parameter setting tabs: General Settings, Font Settings, Image Compress, and PS (EPS) Settings. General Settings options are shown in Fig.4-11.

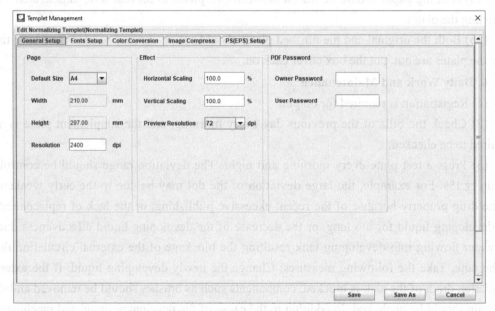

Fig.4-11 General Settings of Normalized Template

1. Page Setup Box

Set the specifications for the normalized page. This setting only can be applied to EPS, PS, and PDF files without page descriptions instead of files containing this information.

(1) Default Size

In the drop-down menu, the corresponding dimensions can be selected, such as A3 and A4. Page dimensions can also be customized " —" Select the "Custom" option to enter the corresponding dimensions in mm in "Length" and "Width" below.

(2) Resolution

This parameter defines the normalized resolution. The standard resolution will be used if the PostScript encoding to be processed does not contain any resolution information. The resolution mainly defines the screen ruling of PostScript shadows. It will not affect the image or font.

2. Effect Setting Box

Set the zoom ratio of the normalized page and the resolution for generating the preview graph.

(1) Zoom horizontally

Sets the multiplier at which the page is zoomed horizontally. The file is magnified or reduced according to the set zoom when normalizing. The default value is 100%.

(2) Zoom vertically

Sets the multiplier at which the page is zoomed vertically, with the default value being 100%.

(3) Preview map resolution

After setting the normalized page, a preview image resolution is generated, which is 72 dpi by default. Resolution can be selected in the drop-down menu, such as 72 dpi, 144 dpi, etc.

3. PDF Password Setting Box

This is only for PDF files with passwords. A password is set in some PDF files without which the file cannot be processed properly. Therefore, the required password must be set here in order to normalize regularly.

(1) Password

Enter the password for opening the PDF file.

(2) License password

Enter the License password for the PDF file.

II. Screening Parameter Setting of the Direct-to-plate Process

Screening parameter setting is accomplished by the output template of the production output process. The screening parameter setting of the KOREJIA process software is achieved by the RIP setting of the CTP output template (see Fig.4-12). The CTP output template outputs the file interpretation to the CTP follower, which is a core part of the process. The template contains five parameter setting tab pages, which are RIP setup, output setup, color manage, calibration and plate setup.

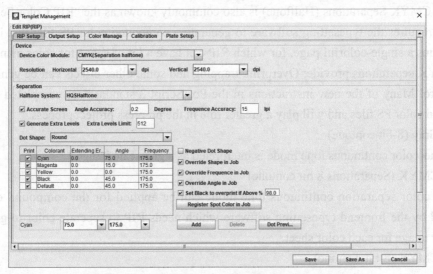

Figure 4-12　RIP Setting—Realization Screening Setting

1. General Settings

(1) Color Space

The "Color Space" reflects the way the output device processes color and image data. Its essence is to tell the users how to process the file to be output, whether RIP color separation is required or whether it is to output binary screening data or continuous tone of data, etc.

A different Color Space can be selected in the "Color Space" drop-down list. Only Gray (Halftone) and CMYK Separations (Halftone) are available for mono-color devices such as imagesetters and CTP machine; CMYK (Halftone) is optional for color inkjet printers; almost all items are available for TIFF output.

① Gray (Halftone)

It represents a single color screening mode used for the front end color separation of the platemaking software. RIP color separation is not required to generate binary latex data.

② CMYK Separations (Halftone)

It indicates RIP color separation screening mode, and the compound color file generated by the front end platemaking software requires RIP color separation screening to generate a binary dot matrix image for each color sheet.

For the "Frontend Color Separation" mode and "RIP Color Separation" mode:

For RIP, all imagesetters and CTP machines are single-color halftone devices, so there are only two following options are available in the color space when the device is an imagesetter and CTP machines.

Gray (Halftone), also known as the "Frontend Color Separation" mode, requires the front end color separation of typesetting software. In other words, "Color Separation" is required in the typesetting software. The generated PS file already contains four single-color pages (assuming only four colors of CMYK). The color data in the four-color plates have been confirmed, and effects such as Overprint and Trapping between the color plates have been realized. So RIP can only explain it and can no longer affect its colors, overprint, trapping, etc.

The CMYK Separations (Halftone) is also commonly known as the "RIP Color Separation" mode. It requires the typesetting software to generate a compound color PS file, that is, the PS file contains a single-colorful page, for which RIP can have a lot of space for color management; The CMYK seperations provide; Overprint, trapping and some other effects can be generated by RIP control Many of the new instructions in the PostScript 3 standard are completed based on compound color PS files and will play a greater role in the prepress project process.

③ Gray (8-bit contone)

Mono-color continuous tone mode is interpreted as a grayscale image.

④ CMYK (Separations 8 bit contone)

RIP color separation continuous tone mode can be applied for the compound color file generated by the frontend typesetting software which needs RIP to seperate color to generate a grayscale image for each color sheet.

⑤ CMYK (8-bit contone)

RIP color separation continuous tone mode can be applied for the compound color file generated by the front-end typesetting software which needs RIP to color seperate to generate a colorful image.

⑥ RGB (8 bit contone)

Output RGB color image mode, like the previous item, which can be applied for the compound color file generated by the front end typesetting software, requires a conversion of RIP to RGB mode and a colorful image generated for each color format.

(2) Resolution

The resolution of the output file is set by the backend output devices. You can also customize the output resolution by selecting the appropriate horizontal and vertical resolution from the drop-down menu. RIP will generate a latex file according to the selected resolution. The output size will be deformed when the selected resolution is inconsistent with the actual resolution of the output device.

2. Color Separation Settings

Whether the parameters of this setting are available or not depends on the color separation method set in "Color Space". In some modes, the color separation parameters are not valid.

(1) Screening Mode

Six AM screening methods are listed in the drop-down list such as "Skew Halftone", "HQS Halftone", "Balance Halftone", "External Halftone", "Normal Halftone" and "LessIntaglio Halftone".

The different screening methods cause different control algorithms of screening accuracy, dot rose spot, and grayscale level "Skew Halftone", "HQS Halftone" and "Balance Halftone" are the most commonly used screening methods. "External Halftone" is a customized dot screening method, which is mostly used for screening of gravure platemaking.

(2) Use Precision Screening

When selected, the users can set the "Mesh Angle Accuracy" and "Dot Frequency Accuracy".

① Mesh Angle Accuracy

It refers to the allowable angle difference between the actual dot angle after RIP screening and the set dot angle. The default value for the system is 0.2°.

② Dot Frequency Accuracy

It refers to the allowed frequency difference between the actual dot frequency after RIP screening and the set dot frequency. The system default is 15 lines per inch.

Users shall not change the above two settings on their own, otherwise, it may cause a collision with the dot.

(3) Generate Additional Tone

When checked, the "Low Limit of Adding Tone" can be set. The default value is 512. The input is the lower limit of the grayscale fraction. The higher the grayscale fraction, the smoother the hierarchy of the output image, especially for the gradient. But it has a certain effect on the explanatory speed and the shape of the dot.

(4) Dot Shape

31 different dot shapes are listed in the drop-down list which can be selected as needed,

and dot shapes can be previewed at different dot angles and screen ruling by clicking the "Dot Preview" button below the template.

When screening way is "External Halftone", the dot shape will be loaded into the "Custom Dot" resource stored in the resource library, and one of them can be selected for screening.

① FM dot size: Set dot size when screening mode is FM screening ;

② Positive-negative inversion of dot shape: Generally, it is only applied to select this output when the image file itself is a negative picture;

③ Dot shape in the overwriting project: selected by default;

④ Dot frequency in overwriting project: selected by default;

⑤ Dot angle in the overwriting project: selected by default;

The ③④⑤ three items above indicate whether to replace the setting in the file to be output with the setting values of the relevant items in RIP. The selection indicates replacement.

⑥ Black overprinting above 95%: Checked by default. It is only valid for RIP color separation.

Overprint is performed when black reaches the set value or more to avoid the difficulty of overprinting in printing and avoid the appearance of white edges.

3. List of Color Printing Settings

The lower part of the template is the "Color Printing Setting List", which is used to set the output of the color plate and the mesh corner net number of lines used when outputting the color plate. Users can select one of the color formats. Printing parameters of each color plate are set through the drop-down menu.

The Default color plate represents all color plates except the color plate listed in the list. If spot color is not added to the template, it refers to all spot color plates except CMYK four colors. You can control the parameters and output of all spot color plate screening through this setting.

At the bottom right of the list are the "Add" and "Delete" buttons. A spot color can be added or removed from the list through these two buttons. After clicking "Add", a line of data will automatically appear in the "Color Printing Setting List". Check this line of data added to change the name of the spot color in the text box below and set the default printing parameters for the spot color in the drop-down menu.

"The Register Current Project Spot Color" option is the spot color plate included in the registration current project. It is set only for the current project. After clicking, the process will register the spot color plate included in the current project into the current template.

If the screening parameter of a spot color is set separately, the spot color name added to the "Color Printing Setting List" must be consistent with the spot color name shown in the "File Information" dialog box of the file after normalization; otherwise, the setting of spot color will not work.

"The Color Printing Setting List" and "Black Overprinting Above 95%" only play a role when separating colors. For PS files where the front end has been separated, the output of a color plate or not and the overprinting relationship between them have been described in the source file, which cannot be changed in the process.

Chapter 4 Output of Printing Plate

4. Spot Color Conversion Setting Items

Spot color conversion is the setting of the processing method of the spot color plate in the project. The spot color conversion processing option is to set the mode in which the spot color plate is outputted, which is divided into the "Conversion of Spot Color to CMYK Output" and "Processing Spot Color by Movable Part Setting".

(1) Turn the spot color to CMYK output

This parameter indicates that all spot colors set for output in the template are converted to four colors for output.

(2) Set spot color according to movable part setting

This parameter indicates that spot color will output according to the properties of each file in the project, that is, the set of spot colors in the "File Information" dialog box. Now, the template just controls whether the spot color is output or not.

Section 3   Solutions to Faults Arising in the Process of Plate Output

I. Image Quality Issues and Its Processing Solutions

In case of quality failure on the plate during outputting the plate, the reason for this failure are as follows:

Computer: The output plate does not match the compatibility of the computer or there are viruses on the computer;

Platemaking machine: Deviation of focal length and power mismatch; laser uncontrollable due to glitches during transmission or on the drum; lack of oil in the screw rail; improper mechanism adjustment.

Processor: Developing temperature and speed settings are unreasonable; The equipment does not work properly due to the failure of maintenance and cleaning; The pressure of each roll shaft and brush is not adjusted properly; The matching ratio of replenishment fluid is unreasonable.

Plate material: The plate material is not ideal in terms of hole sealing; The photoconductor is not evenly coated; The flatness of the plate material is excessive and the edges are irregular, which is not suitable for the platemaking machine.

Table 4-2 shows the quality issues and solutions for printing plate images.

Table 4-2 Quality Issues and Processing Methods of Printing Plates Image

	Image Quality Issues	Reason	Solutions
1	Inaccurate Arrangement	1. Left-right Inaccurate (1) Inaccurate positioning on the left side of the plate, insufficient pressure on the rollers on the side gauge (2) Inaccurate positioning of the scanning platform and loose screw motor coupling (3) For type C machines, the gap between the side gauge nut and screw may be too large 2. Back-forth Inaccurate or inclined (1) Excessive conveying resistance during plate feedings, such as static electricity on the typesetting table or serious static electricity on the plate material (2) In the case of the plate set, check the difference in casing accuracy between plate material 650mm×550mm and plate 1030mm×800mm	1. Adjust the pressure or replace it, refer to *Integrated Commissioning Process* 2. Fasten the fixing screw 3. Verify that the gap between the side gauge nut and screw on the side is less than 0.01 mm (1) Wipe the guide with alcohol (2) Whether the guide plate of the plate setter is consistent with the plate feeding height of the hosts. If the small plate set is free from failure but the large plate is broken, and the friction resistance of the plate setter on the side is too large, contact the Customer Service Department of CRON
2	The layout is unkempt with ash	1. Equipment failure (1) Improper focal length (2) Unsuitable power 2. Fault of processor 3. Unsuitable developing temperature and time 4. Processor is in the improper working condition 5. Plate material's problem (1) Unsuitable plate material sensitivity (2) Vnsuitable sealing of the plate material	1. Confirm the conformity of the plate material used by the customer (1) Check the actual thickness of the plate material and focus the chart to determine the optimal focal length and appropriate power 2. Confirmation of processor maintenance (1) Check the actual development temperature and time (2) Check the working pressure of the water cycle, the water must be sprayed directly, and the pressure of the brush is not equal 3. Check the brand, production date of the plate 4. Contact the plate material supplier
3	The image fails to be sent, the scanning platform will be reset to prompt for emptiness	Data can't keep up, data format is destroyed: (1) The computer is working with other processes being transferred in, such as data transmission (2) Computer Infection with Virus	1. Don't do anything unrelated to work 2 Eliminate the virus
4	Dark lines and dark dots appear in local images	1. The scratches on the back of the plate material 2. Dust is in the corresponding position on the light drum	1. Check the scratches and deformation on the back of the used plate material to confirm whether it is original or generated during the transportation process, find out the corresponding position, and remove the feathers 2. Locate the corresponding position on the light drum and remove the feathers

Chapter 4 Output of Printing Plate

(continued table)

	Image Quality Issues	Reason	Solutions
5	Thin and regular white lines appear on the plate; white lines appear on the field in the image, or black lines appear on the blank part	The laser cannot be controlled effectively, and the corresponding laser drive plate or connection is in failure	It is possible to set the 0 path and 8 paths to 50 through the LaserAdjust program and test the focal length diagram. First, confirm which laser is uncontrolled. Second, Check the connection of the corresponding number of laser products to eliminate poor contact. Replace the laser drive plate if the connection is free from problems
6	Occasionally on an image, there are separated white lines in a segment	Bad contact with the main control panel	Replace the main control panel
7	Occasionally white lines appear on the image	Misconfiguration of negative pressure driven by laser (Note: Thermal model)	Reconfigure negative pressure
8	3% of the dot is missing	1. the laser power too high 2. Variation in printing temperature and time of the processor	1. Determine reasonable exposure power 2. Check the temperature and time of the processor. Attention should be paid to whether replenishment fluid is supplemented as required and the reasonable ratio of replenishment fluid. It is recommended that the matching ratio of replenishment fluid should be the same as that of developing liquid
9	For 98% of dots, white dots are not obvious and streaked	the laser power too low	Adjust the laser power while ensuring clean
10	The whole page is with patterns or brush marks	The pressure of each rubber roller of the processors is not adjusted in place and the pressure of the brush roller is not adjusted in place	Adjust the pressure of each rubber roller and the pressure of the brush roller
11	There are faint streaks or dirt on the image, which cannot be washed clean	development is not clean, developing liquid fails, or processor condition is improperly set	It is possible to increase the temperature of the liquid or reduce the speed of printing, or change the solution
12	Occasionally misalignment of the image head during publishing	Unstable phase lock: (1) Code tray signal is disturbed (2) Unstable motor movement	Check for effective grounding
13	Occasionally there is local scrambling when publishing, and sometimes it goes beyond the graphic	1. File damaged 2. Code tray signal is disturbed	1. Whether it is shown normal during file preview 2. Incompatibility between the computer motherboards and memory can also cause files to be damaged. Replace a computer that is free from viruses first 3. Test to determine that the file is not destroyed during production 4. Check for effective grounding
14	In an image, a full plate of a small white or black dot	Issue of plate material	Test with another brand of plate material

(continued table)

	Image Quality Issues	Reason	Solutions
15	Images marked, caused by non-laser scanning	1. Plate material problem 2. The rubber roller of the processor expands	1. Check whether there is friction during input or output to cause traces of the layout because the plate's film is tender. 2. Check whether each system of a processor damaged to the layout
16	Plate with black lines, divided into irregular black lines and black lines with intervals, 4 mm and 5 mm	1. Oil shortage of guide rail and screw 2. The scanning platform deviates from the straightness of the light drum 3. screw local movement resistance varies	1. Re-clean the screw, front and back guide rails as required, and add grease 2. Check with a clock gauge, the straightness of the light drum and the light scanning platform is less than 0.006 mm; if there is an excess, the parallelism should be adjusted 3. Check the motion accuracy of the screw with a micrometer, it should be less than 0.004 mm, re-polish the screw if it exceeds

II. Other Abnormal Situations and Handling Methods

Table 4-3 shows the system operation abnormal and its handling methods.

Table 4-3 System Operation Anomalies and Processing Methods

No.	Abnormalities	Cause Analysis	Treatment method
1	Unsmooth plate feeding	1. unsuitable vacuum adsorption force for import (D/E type) 2. Unsuitable pressure of input wheel (type C) 3. plate setter does not match the left and right positions of the host 4. Static electricity is found on the plate material or guide	1. Adjust the vacuum adsorption, as shown in the *Integrated Commissioning Process* 2. Execute the LaBoo program and run the test procedure 151 commands to adjust the pressure of the two sets of wheels to be consistent. The adjustment is based on a plate material of 0.15 mm thickness; when the plate is sent to the head, resistance is encountered, and the middle of the plate material is not crooked 3. The guide of plate feeding of the plate setter is consistent with the input rail of the host computer and is centered left and right. There is a distance of more than 3 mm between the left edge of the plate and the side gauge pointer on the side when the plate is sent to the host 4. Electrostatic should be eliminated. The guide can be washed with alcohol, because the static electricity on the plate material which needs to be eliminated
2	Plate unloading is not smooth	1. Insufficient open angle for the plate unloading head 2. The plate base of the thin plate used is soft	1. Adjust the opening angle 2. Liaise with the company's Customer Service Department

Chapter 4 Output of Printing Plate

(continued table)

No.	Abnormalities	Cause Analysis	Treatment method
3	Stuck plate during return	1. It is possible that the plate material has been inclined at plate load so it hit the side gauge when resigning 2. For the type C machine, there is a problem with the installation of the return guide plate	1. Check whether the delivery power and delivery is inclination 2. Return guide plate must be parallel to the input guide plate
4	Tail adjustment failed	1. Inaccurate position positioning of light drum tail 2. Insufficient tail angle when opening, failure to effectively disconnect the lock gear 3. tail left and right torque damage 4. tail rocker arm spring damage	1. Check for the accuracy of parameter 3 2. Check whether the stepping motor, driver and sensor for tail switching are operating normally; Parameter 5 can be adjusted 3. Replace the tail left and right torque 4. Change the tail rocker arm spring
5	The tail slope exceeds standard	1. The plate material is irregular 2. Inappropriate setting of parameter 0	1. Check the angle ruler of the plate material to confirm that the plate material meets the standard; within 200 mm of the middle section, the allowable error is ± 4 figures, about ± 0.4 mm 2. Re-check parameter 0
6	The plate material length fails to match the template	1. The actual length of the plate material is not standard 2. Inappropriate setting of parameter 8	1. Check the actual length of the plate material with an allowable error of ± 10 figures, about ± 1 mm 2. Reset parameter 8
7	Please execute the command of plate unload in the report, there is "no plate on the light drum" or "there is a plate on the light drum"	The detection probe is improperly adjusted with or without a plate on the drum	Execute command 105 to reset the "Probe Detection without a Plate"
8	Anomalous light drum speed during exposure	1. Computer virus attack the programs 2. Light drum belt is loose, or the light drum belt is seriously worn 3. Unstable power supply quality	1. Clear the virus 2. Adjust the tension of the belt 3. Ensure that the quality of the power supply is within 220AC (1± 15%)
9	the temperature of laser box out of range (When the ambient temperature is met)	1. temperature sensor (18B20) damage 2. Constant temperature drive plate may be damaged or have poor wire end contact	1. Confirm that the temperature sensor is operating and can be interchanged with the rack temperature sensor to find the cause 2. Measure the 4P pin of the constant temperature drive board with a multimeter. In normal conditions, the input voltage is 15V and the output voltage is about 9V. If it is always too high or too low, it should be recalibrated

Computer to Plate of Lithographic

(continued table)

No.	Abnormalities	Cause Analysis	Treatment method
10	Failed to lock light	1. Emission light misalignment in the center of the energy detection sensor probe 2. The lens, close-arrange, and energy detection sensor have dust on the surface 3. Bad contact with the end of the wire 4. laser drive plate failed 5. Poor coupling of fiber and laser 6. Fiber damage 7. Laser damage	1. Check the accuracy 2. Clean the dust on each surface 3. Check the voltage of the laser for the correctness 4. Check the drive board for problems 5. Check the coupling of the fiber for light leakage 6. Check the close-arranged fiber for light leakage 7. Replace the laser It can be determined that which part is defective by using the interchange method
11	Balance block adjustment failed	1. Whether the plate materials are standard specifications 2. Whether the motor, drive, and sensor are working properly 3. Whether the sensors is installed properly 4. Whether the parameters are correct, the positioning and slide gears are opened correctly 5. Whether the fixing of each part of the mechanism is accurate	1. Confirm that the plate is standard, without cutting after platemaking 2. Ensure that the motor, drive, and sensor are operating properly 3. Verify proper reset 4. Accurate positioning of the light drum enables the lock gear to be opened effectively 5. Verify the normalization of the movement of each component

Chapter 5 Development of Printing Plate

Objectives:

1. Judge the development condition and set the development parameters.
2. Test the acid-base (pH), temperature, and conductivity of the developing liquid with a detection instrument;
3. Maintain printing equipment such as a developer.

Section 1 Development of CTP Printing Plate

I. Basic Operation of Printing Plate Development

The basic operations of printing plate development are as follows: When the machine is powered on, the necessary process parameters such as development temperature, motor speed, plate size, etc. should be set via the operation interface. After the process parameter setting is finished, select the corresponding program to put the machine in operation. The machine enters the heating state and at the same time, carefully watch the working of the circulating pump by observing the flow state of the liquid or whether the users should the temperature of development has increased with a thermometer.

When the replenishment fluid is finished but the new replenishment fluid bucket is not replaced in time will lead to a low developing tank potion. After replacing the new replenishment fluid, the Dev-Rep can be used in the manual operation to suck the liquid to ensure that the developing tank is spillover. There is an inductive failure in the processor channel water level sensor. If the "Water tank is not full" error is still reported after being filled with water, touch the liquid level sensor with a tool. If other programs are selected by mistake, you will be prompted to change or not, press "NO". Press "ESC" on the new interface that appears to return to the home page.

The following points should be noted.

(1) Emergency stop: Press "Emergency Stop" to stop the machine, pull it back, press the green button, and then press the "Home" button at the corner of the interface to return to the home page interface.

(2) Plate stuck handling: Stop urgently, and then loosen the components such as the rubber roller and then take them out.

(3) If the plate stagnates on the platform in front for a while, it will stop halfway after pushing. At this time the machine will automatically report an error, accompanied by a quick warning. Press the "Home Page" button on the error report interface shortly and then put the finger in the front of the sensor to let the machine operate again. This case must be treated immediately because the plate cannot be stuck in the developing tank for too long.

(4) Bolts must be tightened.

II. Precautions

(1) Confirm the condition of the equipment. Step 9 can be carried out directly after filling in the Table 5-1 for new equipment that has not been used, otherwise release the medicine for cleaning.

(2) Confirm the model and specifications of the customer's CTP equipment and other related information, and fill in the form in Table 5-1.

(3) Drain the developing tank (washing tank if present) and the glue slot and close the valve. Drain the filtered water and remove the filter element. Remove the rubber rollers and brushes for loading, developing, washing and gluing of the processor brushes to be cleaned. Note the order and fill all three channels with clean water and wash them at least twice The middle should be washed and rubbed with a rag or other cleaning articles, and be careful not to pour the water to the circuit part and cause a short circuit.

(4) Fill the developing tank with water and mix developer detergent in proportion and mix it evenly for more than 2 hours (the temperature is set at 30℃).

(5) Wash the removed rubber rollers directly with the developer detergent raw liquid, and wash the rubber rollers until the original color of the original rubber is shown. The both ends of whole rubber rollers are more difficult to clean but must also be cleaned.

(6) The developer brush must also be cleaned if there is polymer stickiness, which can be soaked with developing liquid, soaked with detergent or washed with the original solution of developer detergent with a brush in place with protection measures. However, it requires a developing liquid to neutralize the acid after cleaning.

(7) The detergent will be released after the soaking time of developing tank is finished, and the developing tank will be filled with clear water more than twice for washing and rubbing with a rag or other cleaning supplies in the middle. Efforts should be made to scrub the unitary part

Chapter 5 Development of Printing Plate

and don't pour the water to the circuit part in case of a short circuit.

(8) Drain all the water of cleaning equipment in the developer, install the rubber roller and brush, and pour it into the developing liquid.

At this time, attention should be paid to the effect of water remaining in the developer on developing liquid. The two ends of the circulation circuit should be closed first when the liquid is applied. Open one end of the circulation circuit when a small amount of developing liquid is filled into the developing tank, and use the developing liquid in the tank to drain the water in the pipeline into the filter. Then close one end of the water that has been photographed and open the other end for the same operation. After the water is drained, remove the filter and install the filter element.

The temperature can be increased to adjust the development conditions after the completion of the liquid filled into developing tank, then prepare a test and fill in the test results in Table 5-1.

Table 5-1 Printing Plate Test Record

Company full-name		
Platemaking		
Platemaking machine	Model:	Quantity:
	Mode of plate load:	Exposure mode:
Conditions of use	Energy value:	Rotate Speed:
Plate material used	Model:	Batch No.:
	Specifications:	Average monthly usage:
Development		
Developer	Model:	Capacity:
Developing Liquid	Model:	Batch No.:
	Printed edition:	Usage time:
	Amount of liquid supply:	
Development condition	Temperature:	Speed:
	Dynamic supplementation:	Static supplementation:
Protective glue	Brand:	Proportion:
Baking glue	Brand:	Proportion:
Usage		

I. Fault reflection
II. Insight
III. Fault response

Section 2 Fault Resolution to Development of CTP Printing Plate

I. Installation and Debugging Instructions for Conductivity Sensors and Control Panel

1. Single Processor Parts

Requirements for single processor accessories are shown in Table 5-2.

Table 5-2 Processor Parts

No.	Parts Name	Quantity
1	Conductivity meter	1
2	Conductivity meter bracket 1	1
3	Conductivity meter mounting nuts	2
4	Conductivity control panel	1
5	The overlapping line for the conductivity control board	1
6	Relay	2
7	Protective ring	1
8	(12.9 S) Inner hexagon screw M3* 16	2
9	(304) Flat-head crossed discal screw M4 * 12	1
10	Jam nut M4	1

2. Installation Instructions

Replace the original liquid level instrument and temperature sensor bracket on a processor with the conductivity instrument bracket. Please refer to Fig.5-1 to Fig.5-15 for installation steps and requirements.

(1) The processor's power supply must be turned off, and then the electrical box cover and back cover can be removed (as shown in Fig.5-1).

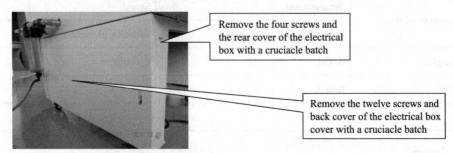

Fig.5-1 Remove the Electrical Box Cover and Back Cover

(2) Turn on the processor's power supply and check the normal lamp of the processor

Chapter 5 Development of Printing Plate

topping-up pump when it is not in the makeup state. Then enter the "Manual Operation" function to make forced replenishment, and then learn about the relays controlling the topping-up pump according to the indicator light to determine the normal conditions of the makeup pump and the power supply line (Fig.5-2).

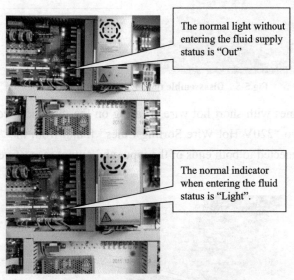

Fig.5-2　Indicator in Liquid Supply Status

(3) Turn off the processor's power supply again. Find the wire in "Conductivity Control Board Casing" to connect the "220V Hot Wire on the Processor Plate Board" and the "24V Power Wire on the Conductivity Control Board", and let it penetrate from the processor wire slot to the installation position of the Conductivity Sensor and relay so as to connect with the control wire of the conductivity control board (Fig.5-3, Fig.5-4 and Fig.5-15).

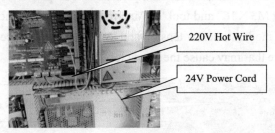

Fig.5-3　Position of Hot Wire and Power Supply

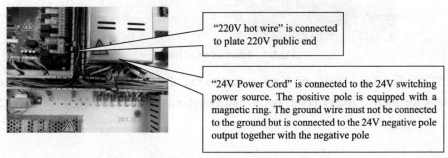

Fig.5-4　Connection of Conductivity Control Board

(4) Remove the right cover of the processor (Fig.5-5).

Remove the right cover of processor with a cross-shaped screw, 4 screws in total

Fig.5-5 Disassemble the Right Cover of Processor

(5) Processor comes with short hot wire shearing on the three-core connector on the liquid supply switch, and two "220V Hot Wire Sorting Lines" found from the "Conductivity Control Board Sleeve" are connected to both ends of the input and output respectively (Fig.5-6 and Fig.5-15).

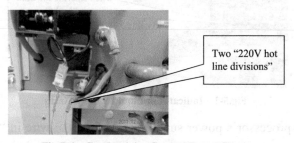

Two "220V hot line divisions"

Fig.5-6 Conductivity Control Board Sleeve

(6) Relay installation: Drill two M3 threaded holes for fixing the relay with the tools "Bosch Torque Drill", "Drill Bit $\Phi 2.5$" and "Tap M3" which are about 5 mm in depth and about 20 mm in spacing. Do not penetrate the processor liquor tank side panel. Then use the accessories "(12.9S) hexagon screw M3×16" and tool "2.5 internal hexagonal wrenches" to fasten the relay. Care should be taken to moderate screw tension and avoid too much force because the liquor tank side panel is a PVC plate that may cause the screw lose (Fig.5-7).

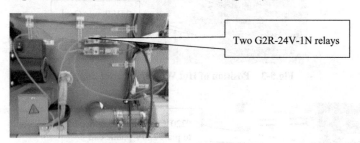

Two G2R-24V-1N relays

Fig.5-7 Installation of Relay

(7) "220V Fire-Line Division" is connected to the "Public End" and "Normal-closed End" of the relays for internal and external control and switching respectively (Fig.5-8 and Fig.5-15).

Chapter 5 Development of Printing Plate

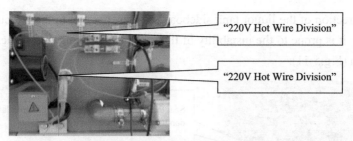

Fig.5-8 "Public End" and "Normally Closed Ends" of Relays

(8) Installation of industrial control relay output line: Find an output power line 1 for "K1 Internal/External Control Switching" and output power line 2 for "K2 Liquid Replenishment Switch" from the "Conductivity Control Board Set", and connect them to the two relays respectively. The other end of the conductivity control panel is connected to the control cable (Fig.5-9 and Fig.5-15).

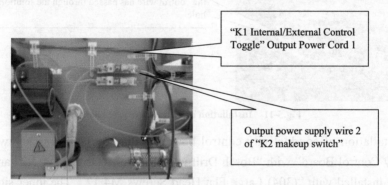

Fig.5-9 Installation of Industrial Control Relay Output Line

(9) Installation of Conductivity Bracket 1: Remove the bracket originally for temperature sensor and liquid level instrument and replace it with "Conductivity Bracket 1" (CTP-GZ-157A) with cross-screws, solid wrench 7mm×9mm, and shifting spanner 6 inches. For the disassembly of a liquid level instrument, the connecting wire must be pulled out from the connecting terminal and then from the hole of liquor tank side plate to be removed (Fig.5-10).

Fig.5-10 Installation of "Conductivity Bracket 1"

(10) Installation of Conductivity Sensor: The Conductivity Sensor is installed on the bracket with the "Conductivity Instrument Mounting Nut", and the conductivity sensing ball must be positioned below the liquid level. Note that the direction of the sensor sensing hole must be

vertical to the direction of the liquor tank sideboard, that is, parallel to the direction of the rubber roller, otherwise, it is prone to the instability of the sensor monitoring due to the frequent flow of the drug solution (Fig.5-11).

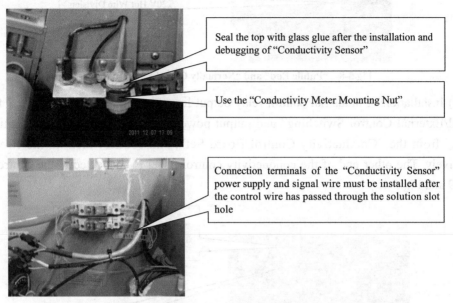

Fig.5-11   Installation of Conductivity Sensor

(11) Installation of conductivity control board: Drill the suspension screw hole M4 of "Conductivity Control Board" with "Bosch Drill with Torque", "Drill Bit Φ3.2" and "Tap M4". The parts are installed with "(304) Large Flat Head Screws M4 12". The inner side is fastened with "Jam Nut M4". Use "Bosch Torque Drill", "Drill Bit Φ3.2" and "Drill Bit Φ13.5" to drill the holes for the "Conductivity Control Board". Protect the conductivity control wire with a "Protective Ring" to install inside the hole (Figure 5-12).

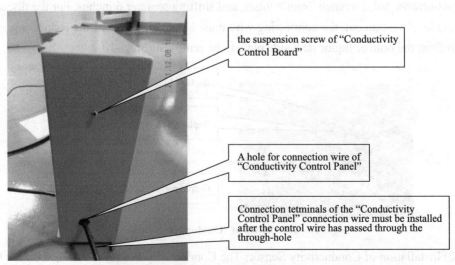

Fig.5-12   Installation of Conductivity Control Board

Chapter 5 Development of Printing Plate

(12) Connection of the connecting wire of the "Conductivity Control Board" to each supply wire, control wire and signal wire: Pay attention to the fixing of each wire (as shown in Fig.5-13 and Fig.5-15).

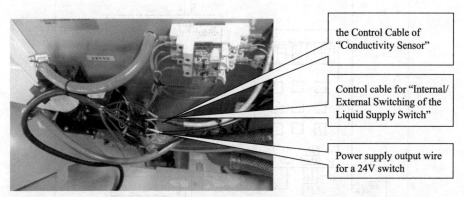

Fig.5-13   Connection of Conductivity Control Board

(13) Installation of "Conductivity Control Panel" and fixing of shell: Restore the electrical cover and outer cover of the processor (Fig.5-14).

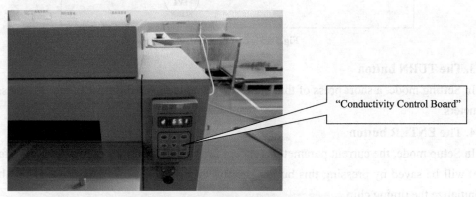

Fig.5-14   Installation of "Conductivity Control Board" and Fixing of Cover

II. Printing Plate Development Parameter Setting

In the display interface of the conductivity control display panel, the first bit is the parameter item, and the last four bits are the parameter value.

1. The "Menu" button

In Display mode, press this button shortly to enter the setting interface while in Setup mode, press this button shortly to toggle to the next parameter setting interface. In any mode, the users can press and hold the button to toggle to the oxidation compensation setting interface.

2. The UP button

Press this button shortly to toggle to the previous parameter setting interface when the screen in Set mode does not flash. A short press of this button will increase the value of this bit when the screen position is flashing. In any mode, press this button long to clear the number of liquid supply times below the lower limit.

107

Computer to Plate of Lithographic

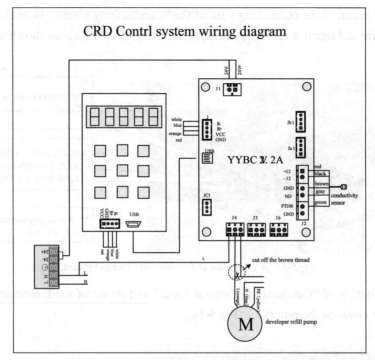

Fig.5-15   Circuit Control

3. The TURN button

In Setting mode, a short press of this button will toggle the flashing bit to modify the setting parameters.

4. The ENTER button

In Setup mode, the current parameters (except for special functions such as restoring factory value) will be saved by pressing this button shortly. In any mode, long key-press of this button will initialize the timing chip.

5. The DOWN button

In Set mode, when this bit does not flash, a short press of this button will toggle to the next parameter setting interface. A short press of this button will decrease the value of this bit when it is flashing. In any mode, press and hold the button to toggle the alarm function on and off.

6. The ESC button

In the setting mode, short pressing this key will exit the setting mode and display the wait interface, and the conductivity interface will be displayed after five seconds.

7. The F1 button

In the rehydration state, short pressing the key will close the current rehydration state. In any mode, pressing the key for three seconds will enable users to enter the forced rehydration state until the conductivity reaches the set point off.

8. The F2 button

In any mode, pressing the key for three seconds will help to switch the rehydration control mode (CRD / punch).

Chapter 6 Quality Inspection of Printing Plates

 Objectives:

1. Be able to discover the scratches, creases, dirty marks and other defects of the printing plate in time;

2. Be able to check the plate dimensions against counterparts, including whether words are lost, scrambling code, incomplete drawings and deformations, etc.

3. Understand the principles of printing plate measurement instruments;

4. Be able to measure the dot and angle of the printing plate with a measuring instrument;

5. Be able to detect the clarity and completeness of screening text and lines;

6. Be able to check the shape completeness and enlargement value of dot with the help of measuring instruments and measurement and control strip, and propose improvements for platemaking process;

7. Be able to conduct a comprehensive inspection of the quality of the printing plate and propose solutions to the problems generated.

Section 1 Common Quality Issues with CTP

I. Detection of Printing Plate in CTP

① Center the image; ② Leading edge is consistent with printing; ③ The width of the tail edge should reach 5 ~ 6 mm, which can ensure the stability of the printing plate in the high-speed operation of the drum; ④ Dot reduction rate within 1%; ⑤ Printing plate is required to be from the bottom ash; ⑥ Four colors overprinting is required to be within 0.01 mm.

II. Common Printing Plate Defects

The contents of the original file and the generated TIFF file are different in some aspects such as image loss, image distortion, etc. Manual visual inspection is more difficult. Generally, the automatic comparison is carried out through some self-test softwares, such as TellRight and other pre-test software, and self-test is carried out in various aspects such as Arts, Imposition, Rip, etc.

External objects cause printing plate scratches, and creases, which may be generated mainly during packaging, transportation and processing. Processing scratches may be caused by the aging of rubber rollers and brushes, or frequent crystallization. Most of the printing plate damages caused by external objects are irregular.

Errors or vibrations in the laser will also cause damages to the printing sheet, which if regular and usually show a vertical line from head to tail.

Insufficient energy, lack of development, or failure of developing liquid can also cause wounds to the printing sheet, which is mainly reflected in the color of the photographic glue with speckles appearing in the full sheet.

Section 2   Printing Plate Instruments

I. Introduction of Function

The CRON Techkon Spectro Plate-Connect is a plate measurement device that has excellent image capture quality and original graphics algorithm and can accurately read any screening size and screening technology: FM screening, AM screening or hybrid screening. The white illumination spectrum and dynamic color evaluation allow the device to read all types of plates and surface coatings. The advantage of the plate measurement device is not only shown in the printing plate reading. This multifunctional device is equally excellent at performing film dot measurements and CMYK four-color printing paper measurements.

Fig.6-1   the Plate Measurement Device (Spectro Plate)

II. Measurement Dot Type Setting and Measurement Method

According to the type of screening: AM screening and FM screening, select the corresponding setting, put the instrument on the position that required measurement by control bar of the printing plate, press the green measurement button, and you can see the corresponding data on the instrument: the dot area, if it is the AM screening, you can also see the number of the line of the screening and the angle of the screening. Fig.6-2 and Fig.6-3 show the measurement functions of FM screenig and AM screening respectively.

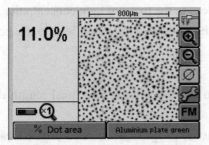
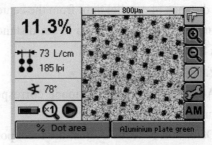

Fig.6-2 FM Screening Measurement Function   Fig.6-3 AM Screeniing Measurment Function

The Techkon Spectro Plate (as shown in Fig.6-4) can be controlled on the PC. The operation interface of the instrument can be shown, and the instrument screenshots can be saved. The Techkon Spector Plate can be connected to a computer and displays the instrument's screen information on the computer when the device is working.

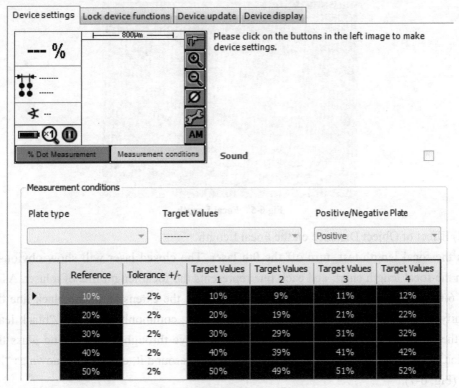

Fig.6-4 Techkon Spectro Plate

Section 3　Printing Plate Output Inspection

I. Determine the Optimal Focal Length of the Plate Material

1. Definition of Focal Length

It is a way to measure the concentration or divergence of light in an optical system that refers to the distance of parallel light from the center of the lens to the focus of light concentration.

In short, the focal length is the distance from the middle point of the lens to the point at which the light can be focused.

2. Definition of Zoom

Adjustments should be made corresponding to the focus point and focal length during exposure.

Definition of focus: Adjust the focal point so that the image to be scanned can be located in the focal length and the imaging is clear.

(1) Normal focal length

A plate of focal length test with CRON is output under the original working power and printing conditions. Put the plate tail near the human body on the table or the platform. The focal lines in the focal length test block should be evenly arranged without streaks, as shown in Fig.6-5.

Fig.6-5　Focal Length

(2) Effect of Object Distance on the Focal Length

In the focal length test, turn off the 0th laser. The closed laser will show obvious black lines in the focal length chart, and each line represents the focal length line of a laser. As shown in Fig. 6-6, when the distance between objects is small, the extension of the black line through a magnifying glass is darker than the surrounding lines, commonly known as "black leakage". When the distance between objects is too large, observe through a magnifying glass that the extension of the black line is lighter than the other lines around it, commonly known as "leakage white" (Fig. 6-7).

Chapter 6 Quality Inspection of Printing Plates

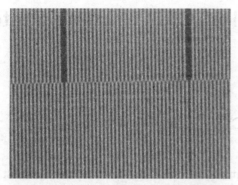
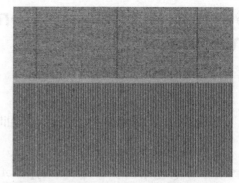

Fig.6-6 Focal Length with Small Object Distance   Fig.6-7 Focal Length with Large Object Distance

(3) Method for Determining the Optimal Focal Length

Method 1: Observe from the right to left on the focal length test image with a magnifying glass. 1×1(pixel×pixel) dot appears in the Mth lattice, and disappears after the Nth lattice. The optimal position of the focal length should be one grid to the left of M+N/2.

Method 2: Select the smallest dot area that can restore the dot. For example, 1% of the dot area can already be restored out of dot. Select the magnifying glass to observe the 1% dot area from right to left. The restoration of the dot can be clearly found in the Mth lattice, and the loss of the dot is serious before the Mth lattice; the dot is still clear at the Nth lattice, but the loss of the dot is obvious after the Nth lattice. In this case, the optimal position of the focus should be one grid to the left of M+N/2.

II. Determine the Best Exposure Conditions for the Plate Material

The optimal exposure and printing conditions required by different brands of plate material are not the same. For example, a black wood UV plate material is suitable for a black wood developer. If the UV plate material manufacturer provides a standard solution ratio of 1:4, developing liquid conductivity of 60ms/cm, brush speed of 100r, and washing temperature of 25℃, then the time of development is 20s.

Then output a variable power test chart. Output a variable power test diagram, through the detection method of dropping acetone, detect whether the acetone solution diffuses on the plate without pictures and images is left with bottom ash: the phenomenon of the diffusion of blue circles is regarded as the bottom gray; the phenomenon of the diffusion of white is regarded as no bottom gray. The dot detector is used to detect the no-bottom gray area, and the dot reduction of 50% dot area is 50% or less than two or three points, while ensuring that there is no bottom gray, the power value of slightly less than 50% dot reduction can be selected as the best power. If the optimal power is large and close to the power limit of the equipment, the optimal power can be reduced by increasing the time and temperature of the developer.

Section 4 Quality Control of CTP Printing Plate Imaging

I. Factors that Influence the Quality of CTP Printing Plate Imaging

1. Plate Material Quality

The imaging quality of different types of plate materials varies. For an enterprise, it is best to choose CTP plate materials of fixed manufacturers and models, so as to improve the stability of platemaking quality.

2. Equipment Performance

The exposure performance of different devices varies, and the illumination and uniformity of the light source per unit area will affect the uniformity of the dot.

3. Developing Conditions

The chemical composition, temperature, and concentration of developing liquid are all key factors affecting the quality of platemaking. Also, after a certain number of plates are developed, part of the resin layer in developing liquid will form much floccule that will be attached to the finished plates. It will cause dirt when printing if it is not addressed.

4. Process Control

It refers to the setting of various process parameters, such as exposure time, development time, etc. In addition, it is best to complete the same plate at one time, so as to ensure the accuracy of overprinting.

5. Environmental Conditions

It refers to the temperature, humidity, and light conditions in the printing workshop. It should be set within the range required by the sheet.

6. Post-treatment process

When some dot sensitive glue residue appears on the printing plate, it can be repaired with a retouching pen.

II. Quality Control Scheme for CTP Printing Plate

The prerequisite for quality control of CTP platemaking is to adjust the platemaking equipment so that the entire CTP system is in the best state.

Exposure and development are the most important processes during CTP, so the adjustment of the platemaking equipment is mainly about the control of the exposure parameters and development process.

(1) Control of Platemaking Machine's Exposure Parameters

To make good use of CTP, it is first necessary to control the exposure parameters of the platemaking machine so that its optical and mechanical systems are in good condition. The sensitivity performance of the plate material must be tested after the user gets a plate material that

Chapter 6 Quality Inspection of Printing Plates

adapts to the platemaking machine exposure mechanism and matches the wavelength range. The test items include laser focal length and focusing testing (FOCUS/ ZOOM TEST), laser luminous power and roller speed testing (LIGHT/ROTATE). Among them, the laser focal length and zoom, power and speed can be tested in combination.

Generally, a platemaking machine comes with its own internal exposure parameter measurement and control strip. It is easy to detect the state of hardware equipment such as whether the exposure of the printing plate is appropriate and the focusing of the laser head is correct through the color blocks or patterns on the measurement and control strip.

(2) Process Control of Processor Development

After the printing plate is exposed normally, normal development is required inside the processor to obtain a simulated image. Therefore, the status of the processor must be tested and monitored.

As the use goes on, the hardware devices will age and there will be some differences between the setting and actual values. The state of the processor must be tested. During the test, the processor's "Actual Temperature" needs to be measured via/by a special developing liquid thermometer. The large measuring cylinders or measuring glasses are used to monitor the "Actual Dynamic Supplement" for the processor. If the setting value differs too much from the actual value, the cycle system needs to be improved or the sensor device should be replaced.

Developing liquid match test is required when the processor hardware status is monitored. Users can use the standard digital printing plate measurement and control strip from major companies for testing. In addition, the self-made printing plate control strip can also be tested. The change of dot on the printing plate can be analyzed by using the digital printing plate control strip, so as to judge whether the printing plate is washing properly (the premise of normal printing plate exposure) and whether the parameters of development temperature and development speed are set correctly. In general, 2%~98% of the dot on a normal CTP printing plate should be sufficient, a 50% dot gain should not exceed 3%, and 95% dot will not be pasted, and there will be no parallel phenomenon.

III. Digital Monitoring by Digital Platemaking Measurement and Control Strip

The CTP system is part of the digital workflow, so a digital control method is essential for quality assurance. The digital platemaking control strip can reasonably and effectively control the imaging quality of the CTP printing plate.

The digital measurement and control strip used for CTP printing plate control mainly includes GATF digital platemaking control strip, Ugra/Fogra digital platemaking control strip, KODA digital printing plate control strip, Heidelberg digital printing plate control strip, etc. Among them, the most widely used and most important are the Ugra/Fogra digital platemaking and GATF digital platemaking control strips.

(1) Ugra/Fogra Control Strip for Digital Platemaking

The control strip contains six function blocks and control areas, as shown in Fig.6-8.

Fig.6-8 Ugra/Fogra Control Strip for Digital Platemaking

① Information area: includes output device name, PS language edition, screen number of lines and dot shapes, etc.

② Resolution block: contains two semi-circular regions. The lines emerge from a point in a radial array; the density of rays is in line with the theoretical resolution of the output device. Form a quarter circle that is more or less, opened or closed, in the center of the line. The smaller and more round they are, the better the quality of focusing and imaging is. The positive line is on the left and the negative line is on the right.

③ Linear block: It consists of horizontal and vertical micro lines that are used to control the resolution of the printing board.

④ Checkerboard area: A checkerboard square unit consisting of $1 \times 1, 2 \times 2, 3 \times 3$, and 4×4 (pixels $\times$ pixels). It controls the resolution of the printing plate to show the difference between the exposure and development technology.

⑤ Visual reference step-wedge (VRS): Control the image transfer of the printing sheet.

⑥ Screen tone step-wedge: It is mainly used to determine the transfer characteristics of the printing plate tone by measurement. The 1%, 2%, 3% and 97%, 98%, and 99% color blocks provided can also be used for visual judgment on the tone that can be ultimately copied in the high-tone and dark-tone area.

Among them, VRS is a special part of the Ugra/Fogra digital platemaking control strip. It is a basic element for image transfer control to control the stability of the printing plate and standardize the production process of the digital printing plates. The VRS contains pairs of thick mesh wire reference blocks, around which is a fine screening region. There are 11 VRS in the control strip and are incremental by 5% from 35%~85% of the dot area. In the case of an ideal state and linear replication, two regions in VRS 4 should visually share the same order tone value. However, the VRS of both regions with the same tone is higher or lower than the VRS 4, depending on the type of printing plate and the selected calibration conditions. The VRS is a very ideal process control piece, which can be used to visually judge the difference from the selected condition without the measurement.

(2) GATF Digital Platemaking Control Strip

GATF digital platemaking control strip as shown in Fig.6-9.

① Information area: Includes output device name, PS language release, screen number of lines, dot shapes, etc.

② Horizontal vertical fine lines of positive and negative graphs: Test the resolution of the system and control the exposure intensity.

③ Checkerboard area: A checkerboard square unit consisting of $1 \times 1, 2 \times 2, 3 \times 3$ and 4×4 (pixels $\times$ pixels).

④ Micron arc area: Includes positive figure and negative figure pattern micron arc area. The smallest set size is used to detect the system with camber segments, and the micron arc pattern is the biggest challenge to the system. If a system maintains favorable details of the arc of positive graph and negative graph, it indicates that the system is in favorable exposure conditions.

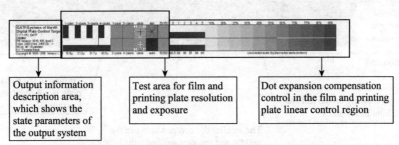

Figure 6-9　GATF Control Strip for Digital Platemaking

⑤ Star object: Test the exposure intensity, resolution, and tone transfer characteristics of the system.

The remaining part is two sets that match the tone step-wedge. The difference between the two tone step-wedges is that the above one avoids the compensation procedure of the RIP applied to other files, while the following one does not skip the compensation setting. A comparison of the two step-wedges indicates the impact of the compensation process. With the tone step-wedge, a magnifying glass is required to observe the highlights of the imaging system and the limitations of dark tone. The tone step-wedge is then measured from 10% to 90% using a densitometer to build the dot gain value curve.

(3) CRON's Digital Platemaking Measurement and Control Strip

CRON standard digital platemaking measurement and control strip generally contains five functional areas, as shown in Fig.6-10.

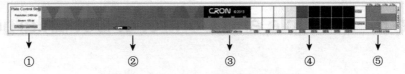

Fig.6-10　Measurement and Control Strip

① Information area: Includes the output device name, PS language version, resolution, number of screen cables, etc.

② Development effect area: See Figure 6-11, which is divided into seven square areas. Through the color contrast between the inverted triangle area in the square and other areas in the square, the square with the closest color is within the 3—5 interval, indicating good development effect. The square with the closest color is in the 1—2 range, which indicates the excessive impact and needs to adjust the development conditions. The square with the closest color is in the 6—7 range, which indicates that the development is insufficient, and it also needs to adjust the development conditions or change the potion.

③ Pixel area: A circular area composed of 1×1, 2×2, 3×3 and 4×4 (pixels×pixels) to detect the effect of laser focus with corresponding accuracy and plate resolution.

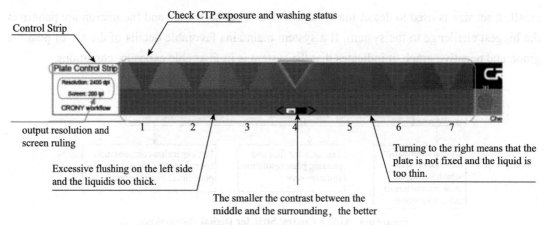

Fig.6-11 Development Effect Area

④ Dot reduction area: Composed of 0%, 1%, 2%, 5% and 50%, 97%, 98% and 99% dot area. It mainly detects the reduction effect of small outlets and large outlets of CTP equipment, and can also be used to make visual judgment of the order that can be copied in the high profile and dark modulation area. This area is divided into upper and lower areas, the following one bypasses the compensation procedure of RIP applied to other files, while the upper one does not bypass the compensation setting. The comparison of the two ladder measures clearly indicates the impact of the compensation procedure.

⑤ Line area: Composed of horizontal and oblique lines of 1×1, 2×2 (pixels×pixels). It can detect the resolution of the printing plate, and can also be used to determine whether the initial exposure point of the image is the best position.

Section 5 Principle and Control of Dot Enlargement

I. Printing Dot and Dot gain

The dot is the basic unit of ink adhesion, which plays a role in transmitting tone and organizational color. Dot gain refers to the gain of the dot printed on the printing material relative to the dot on the color separator. Dot gains damages the print to varying degrees, disrupting the balance of the picture. However, it is impossible to print without the dot enlargement due to technical and light absorption reasons, as shown in Fig.6-12.

One of the goals of printing production control is to set the corresponding dot gain standard for all printing press groups by paper and consider this dot gain standard value when making films, so as to control the dot gain through process compensation to achieve the ideal result of printing image color and tone reproduction. The expansion of the printing dot needs to be measured in order to accurately obtain the dot gain standard.

Chapter 6 Quality Inspection of Printing Plates

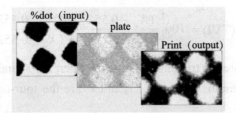

Fig.6-12 The state of Dot Transmission in Printing Process

Dot gain is usually measured by printing several proofs with dot step-wedges and test strips containing ground, 50%, and 75 % dot content (e.g., Brunner test strips) and any image plate under specific printing materials, equipment, and ideal printing pressure conditions. During printing, make sure that each proof dot is neat, real, and free from ghosting deformation. Then, the four-color solid density,50% and 75% area ratios of the printing dot at each proof are measured with a reflective dot densitometer, respectively. Finally, the printing dot is the difference from the 50% or 75% value to measure the dot gain value.

II. Tone Value Increase (TVI)

1. Dot Gain Based on Density Values

In practice, the dot gain should be modified first in any method used to calculate color. Dot gain is the most influential factor in printing color. Typically, the density calculation is used in the calculation of the printing dot area and dot gain quantity, i. e., the Murray-Davies formula is used:

$$a = \frac{1-10^{-D_t}}{1-10^{-D_o}} \tag{6-1}$$

Or use Yule-Nielson formula:

$$a = \frac{1-10^{-D_{t/n}}}{1-10^{-D_{o/n}}} \tag{6-2}$$

Where, a is a unitary primary color ink tone area ratio with a color density of D_t and D_o refers to the printing solid density. Therefore, it is very convenient to control the printing conditions with the density value. Calculate dot gain can be obtained by subtracting the defined value and the result of the calculation by a.

2. Dot Gain Based on Tristimulus Value

As stipulated by ISO/TC10128, the dot gain value (Tone Value Increase, TVI). can be calculated by measuring a series of tristimulus values of color scale in combination with solid four-color tristimulus value.

$$\text{Black and magenta dot gain values} = 100\left(\frac{Y_P - Y_t}{Y_P - Y_S}\right) - \text{TV}_{\text{Input}} \tag{6-3}$$

$$\text{Yellow dot gain value} = 100\left(\frac{Z_P - Z_t}{Z_P - Z_S}\right) - \text{TV}_{\text{Input}} \tag{6-4}$$

$$\text{Cyan dot gain (TVI)} = 100\left[\frac{(X_P - 0.55Z_P) - (X_t - 0.55Z_t)}{(X_P - 0.55Z_P) - (X_S - 0.55Z_S)}\right] - TV_{\text{Input}} \quad (6\text{-}5)$$

Where, X_P, Y_P, and Z_P are the tristimulus value of paper; X_S, Y_S and Z_S are field cyan, magenta and yellow and black tristimulus values; X_t, Y_t and Z_t are the four-color color scale tristimulus value of different tone.

III. Methods to Avoid Dot Enlargement in Platemaking

Dot is a unit that shows the hierarchy, tone and color of printing, and changes in printing dot often lead to quality problems such as color reproduction distortion, tone level reduction, etc. Dot transfer issues common in the printing process include the expansion, sliding, ghost of dot, the most common of which is the dot gain. Two aspects contribute to dot gain: Factors such as the non-linear characteristics of imagesetter or CTP and the tone transmission characteristics of the exposure system; printing press, types of paper and ink, printing pressure and other printing process conditions.

1. Dot gains compensation for the output stage

Ideally, the amount of laser output should be positively proportional to the dot area in different graphic parts of the file after the imagesetter or CTP platemaking machine has received the PS or ONE-BIT-TIFF file. Every 1 pixel of the electronic image is output as 1 dot, and the value of the grayscale of the pixels corresponds to the size of dot one by one. However, this is not the case for most of the actual output results due to process accuracy in machine manufacturing, etc., and tends to be non-linear to a certain level. Together with other factors such as optics and development, it will cause the tone of the dot image output on the software or CTP printing plate to deviate from the original pixel value.

Measures must be taken to correct the deviation so that the pixel value is linear with the final output in order to obtain the required dot size. The correction of this deviation can be achieved using the transform function in the PS page description language. This compensation needs to be carried out before RIP, and the purpose of compensation for dot gain generated during the platemaking stage is achieved by correcting the PS color separation file. Generally, this process is achieved by the following two means:

(1) Compensation through linearization of the platemaking process software

RIP manufacturers provide a customized transform function in their products, i.e., they can invoke the transform function operator of PostScript language in their RIP. The method of dot gain compensation is usually called linearization. Linearization must be made before the new imagesetter is officially put into production; linearization in the current state should also be done after replacing a different developing liquid or film.

For the CTP platemaking machine, accurate dot control is required when changing the printing plate or processing conditions and the FOCUS, ZOOM, or laser power values of platemaking machine, and the linearization under this condition should be redone accordingly. Practices for linearization compensation are shown in the relevant chapters earlier.

(2) Transfer function via Adobe Photoshop software

For the dot gain due to improper imagesetter calibration, it is possible to achieve compensation by establishing a custom transfer function curve using the transfer function built into Adobe Photoshop software in addition to the linearization of the output device.

The specific operation of using transfer function for dot gain compensation in Photoshop is as follows.

① Output the test piece from the imagesetter, and measure the density values of different dot area ratios on the film by using a transmission densitometer;

② Execute the "Page Setup/Transfer Functions" command in the File menu;

③ Calculate the required adjustment values, and enter the calculated values into the corresponding boxes of the "Transfer Function" dialog box. The calculation means that if the output of 50% of the dot is specified, the dot of the imagesetter output is 52% so that the dot gain value of the output is 2%. To compensate for this expanded value, 48% should be entered in the 50% text box of the dialog. As a result, the required 50% dot will be obtained when the color separation file containing this transfer function is called for the output. The setting of the transfer function can vary from color plate to color plate or the same curve can be called by all color plates. If this compensation is used, it should be saved as DCS or EPS format when saving the color separation file, and the "Include Transfer Function" option should be checked, as shown in Fig.6-13.

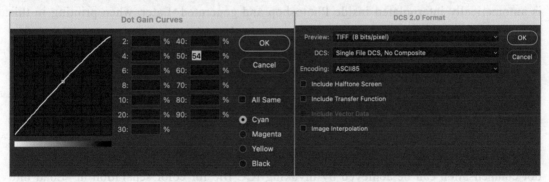

Fig.6-13 "Transfer Function" Dialog Box

It should be noted that Adobe does not encourage users to set the transfer function in Adobe Photoshop software, but rather suggests that users should use the imagesetter's own Calibration Procedure to do this step when outputting it on the phototypesetting machine (see Adobe Systems Incorporated. PostScript Language Reference Manual)

2. Printing Dot gain Compensation

Dot gain in printing refers to the one that is caused by diffusion when the ink is absorbed by paper during printing. The printed dot gain causes the printed dot area to be different from the dot area required, resulting in inaccurate printing color.

The different printing press and paper combinations will have different dot gain values. It is worth mentioning that the compensation for the printed dot gain is not obtained by subtracting the measured value and the theoretical value, but by the dot gain curve, which gets the compensation value according to the inverse function relationship. For example, in dot A at 50%, the dot area obtained after the dot gain is 75% (point C), but it is not actually compensated by 75%-50%= 25% dot area. Instead, a vertical line is made from point B to the Y axis, and point E (30%) on the X axis corresponding to the other intersection point D of the curve is dot area after dot gain compensation.

The dot gain generated by the printing process is generally compensated for color separation in Adobe Photoshop software, and the four independent channels of CMYK or the overall tone of the color image can be set to dot gain compensation, respectively. In addition, Adobe Photoshop software provides grayscale image (converted to grayscale images when color separation) and spot color compensation for dot gain.

(1) Set dot gain compensation via Photoshop color separation

① Make color blocks test files with different dot area ratios in AI; make dot color blocks with a single color of CMYK and color blocks with four color overprinting respectively.

② After printing the test file under given printing conditions, the actual dot area of each color block is measured by a reflection densitometer.

③ Open the test file in Adobe Photoshop, select "Edit" →"Color Settings" →"Custom CMYK" options, select the "Curve" in the "Dot Gain", and enter the measured dot area into the corresponding position in the "Dot Gain Curve" dialog box.

④ The dot gain curve for this printing condition (ink, paper, printing pressure, etc.) is automatically interpolated inside Photoshop and generated. Later, when printing in this condition, the expanded curve of printing dot obtained has the best compensation effect for the current dot gain.

In addition, "Standard" can be selected in "Custom CMYK". The dot gain compensation value that has been set can be called by modifying "Ink Color". Different dot gain values can be invoked using different ink and paper combinations, for example, dot gain is 9% (Eurostandard coated), 15% (Eurostandard uncoated), and 30% (Eurostandard newsprint) when used with European ink standards with coated paper, offset paper, and newsprint, respectively. In addition, dot compensation can be achieved by invoking ICC file for color separation in "Load into CMYK" in addition to customizing dot gain if you make an ICC file under a certain printing strip.

(2) Compensation curve of printing press in platemaking process

The compensation curve of the printing press in the platemaking process is to find the dot gain characteristic of the printing press and generate a compensation curve according to the target dot gain data. For example, the increase rate of 175 lpi square dot at 50% dot is generally controlled at about 15%. Please refer to the relevant chapters above for specific operation methods.

3. Dot Gain Compensation for Text and Graphics

In production, many colors are edited in graphical class software (such as Adobe Illustrator) and typesetting software (such as Founder FantArt). However, none of this software is equipped with a dot gain compensation function, which requires consideration of the expanded color of the printing dot when setting color. The professional practice is to have a printing color pattern (color code) for reference when setting such colors, such as the PANTONE color code.

Chapter 7 Test and Anomaly Handling of Platemaking Equipment

Objectives:

1. Be able to adjust laser energy parameters of the equipment;
2. Be able to judge issues of poor printing ink;
3. Be able to judge and deal with quality issues of printing plates.

Section 1 Printing Plate Output and Test of CTP

CTP printing plate detection is not only an important link of printing plate output, but also the basis of normal output and quality guarantee of printing. Preparations before conducting comprehensive testing are as follows:

(1) Use a multimeter to check: the internal resistance (on the switch) at the input of two groups of AC220V and internal resistance of 24V and 5V of switching power supply to determine that there is no short circuit or open circuit in the circuit;

(2) Whether the cord color line order of each drive is accurate.

(3) Petermine the 0th position of the drum code: The LED indicated by the LP of the code tray is on when the rotation of the drum front gauge is on the same level as the axis of the drum on the screw side and can be measured by using a steel ruler.

(4) See the parameter two servo motors, referring to process documents.

I. Reset Test

The reset test is mainly carried out on the actuator, which is installed in the peripheral direction of the drum. This function checks whether all motors are in the zero position, When the sensor LED light is on, each actuator is in the highest position, and there is a gap of about 0.5mm to the respective hardware limit.

Inspection part are as follows:

(1) Motor sensor for loading/unloading plate on at head and tail;

(2) The motor sensor of a pressing roller for plate loading;

(3) The motor sensor for the pendulum roller rising;

(4) The motor sensor for plate loading and plate blocking;

(5) The driving motor sensor for left-right balance;

If the program commands cannot be finished, there will be corresponding prompts. If you wait for a certain motor or a certain sensor, it is necessary to check the corresponding motor drive and whether the input of the driver is correct and the output is correct; In case of a sensor, check that whether the sensor is valid and installed correctly.

Ensure that the drum does not collide with the mechanism installed along the periphery of the drum when it runs, thus ensuring the safety of the equipment operation.

II. Head Loading Test

It mainly tests the accuracy of head positioning and the correctness of head mechanism control.

III. Head Unloading Test

It mainly tests the accuracy of the plate unload positioning of the drum and the correctness of the plate unload mechanism control.

IV. Tail Test

It mainly tests the accuracy of tail positioning and the correctness of tail mechanism control.

V. Position of Balance Block and Drive Test

It primarily tests the accuracy of balancing block position and the correctness of balancing block control and adjustment.

VI. Test on Plate Feeding Pressure, Drum Vacuum Pressure, Deflation and Dust Pump

It is mainly for the test and adjustment of the press pressure, drum vacuum air pressure, deflation, dust pump.

VII. Side Gauge Mechanism Test

Test the reset accuracy of the side gauge and the side positioning of the plate material.

VIII. Scan Focusing Lens and Object Distance Lens Test

It is mainly for the control and adjustment of the focus lens and objects distance lens on the scanning platform.

Chapter 7 Test and Anomaly Handling of Platemaking Equipment

IX. Plate Loading Test

The plate loading test is to debug how to install the plate material correctly on the light drum, and how to judge whether the plate material installed on the drum meets the working requirements of the equipment.

1. Detection sensor test for there being a plate or not

A reflection sensor is installed in an appropriate position on the drum to detect whether there is a plate on the drum. When load the plate, the equipment will detect whether there is a plate on the drum; if so, it will be prompted in the information column and plate unloading will be required first.

2. Test after loading

(1) Adjust the position of the side gauge

After the plate is installed on the drum, a straight ruler is used to measure whether the distance from the two sides of the plate edge to the edge of the drum is the same. If not, adjust the side guage parameters to ensure that the plate is centered in the drum.

(2) Determine the width of the tail sandwich plate

Draw a straight line at the two ends and middle of the edge of the tail with a pencil, and remove the plate from the drum. The distance from the scribe to the edge of the plate should be measured by using a straight ruler, which is required to be 4-5 mm and the parallelism is less than 0.2 mm. Adjust parameter 9 if it is not accurate.

(3) Accurate positioning of tail position

It is found that the tail position is decreasing gradually and the decreasing average value is 5. In such a case, the decreasing average value should be added to parameter. Conversely, the incremental average value will be subtracted in case of an increment.

3. Plate Inclination Test

Check in the information prompt window, [Tail slope over the limit: ××]. In this case, the change in XX value should be less than ±4.

4. Standard Plate Length Adjustment

Ensure that the board used is standard whose plate length is accurate and is a standard rectangle.

5. Balance Block Adjustment Corresponding to the Drum Position

Measure with a straight ruler, and rotate the drum to make the height of the center of the balance block detection lever consistent with the height of the axis of the drum of the equipment. Now, don't let the drum rotate, and mark the plate with a pencil according to this center level. If the measurement is inconvenient, the wallboard bracing plate can be used as an auxiliary benchmark. For example, for the balance block in the outside of the head, the marking can be made on the drum and the length of the head can be measured by a circle ruler and can be measured in stages. Repeat these steps to measure another balance block and make a mark. If another balance block is on the outsides of the tail, the marking can be made on the drum and the distance can be measured to the tail. It should be noted that if the balance block adjustment is not accurate, the drum will vibrate and sound will be higher when it rotates.

X. Laser Focusing and Energy Test

1. Focal length adjustment

Select the TIF file that is capable of the focal length test, such as the focal length test diagram file "Test focus745X605_2400_90". Set the number of focal length bars on the output plate; According to the set parameters, the output of the focal length test chart work. When placing the plate, put the plate head upward to observe the focal length line with a magnifying glass. If there is a thicker line in the focal length line, it indicates that the object distance is short, and the object distance parameter should be increased, and if the opposite it should be reduced; Find the corresponding dpi focal length value and modify it. The optimal focal length distribution is regulated. The focal line presented by the optimal focal length is uniform thickness. The dots are clear and complete and are symmetrical on both sides of the optimal focal length, and the farther the dot is away from the optimal focal length, the more dots will be lost until the dot cannot be seen.

The best discriminating conduct focusing length is as follows:

Make the tail side to approach the human body and put it on the platform, and the test chart is horizontally distributed with 1% ~ 99% of the dot area. Use a magnifying glass to see the 1% dot area from right to left. 1% of the dot can be observed in column M and marked in this column. It will then be found that the dot becomes increasingly clear and then starts to decrease again. If 1% of the dot in column N disappears, this column will be marked. At this time, the middle region of the two columns is found, which should be the best focal length.

2. Laser Energy Test

Compare with focal length test steps when performing laser energy testing. Select the TIF file that can perform a focal length test project to enter the setting interface of project template, and check "Power Test". Set "minimum power": Every machine has its power range, which requires setting an appropriate starting power; Output power test chart operates according to the set parameters; Selecting the appropriate washing condition to develop.

Select the best energy according to the following items:

(1) The best energy region without bottom ash can be dropped into the blank region with acetone or anhydrous ethanol. The blank area did not change to any bottomless grey. Bottom ash will be found in case of diffuse blue circles.

(2) The optimal energy region dot reduction should be controlled within ± 1%. It can be tested via the dot tester.

(3) 1% of the dot and 99% of the dot in the optimal energy region were presented fully and clearly.

(4) The optimal energy value selected according to the above three points should be within the energy range of CTP and at a certain distance from the energy value of the upper limit. If it is close to the upper limit or exceeds the range, the washing conditions need to be appropriately

Chapter 7 Test and Anomaly Handling of Platemaking Equipment

adjusted until the energy value meets the requirements.

XI. Position Adjustment of Plate Image

Output a flat dot first. In the output position set by the project template parameters, the upper and lower margins are 0; check to be centered horizontally and vertically. Check the following key points:

(1) There shall be no interference streaks in the dot cable; check whether the dot diffusion rate in each region is consistent; If there are dark spots on the layout, this situation is generally because the light drum has dirty spots, resulting in local bumps after the version is loaded into the light drum, and the raised part will produce dark spots because of defocus.

(2) Measure the distance of the head image to the edge of the front plate. Adjust the corresponding parameter if it does not match the design's theoretical value. For the distance between the edge of the head and the exposure point, ten parameters are 1 mm, and a decimal point can be entered;

(3) Measure the distance from the edge of the image to the edge of the plate (horizontally of the drum). In case of discrepancy with the design theoretical value, adjust the corresponding parameter. Record the distance from the exposure point to the project output point of the recording head; the ten parameters are 1 mm, and the decimal point can be imported.

XII. Maintenance of the Linear Motor System

Regular maintenance of linear motor grating can improve the operation accuracy and stability of the laser auxiliary scanning mechanism and ensure the dot quality; when cleaning, the alcohol, acetone and other liquids are strictly prohibited, and only dry cotton cloth can be applied.

XIII Automatic Refueling System Maintenance

Check the lubrication state of the equipment auxiliary scanning mechanism and the oil storage amount of the guide rail. The auxiliavy scanning mechanism is self-lubricating. After the oil is refilled regulary, maintenance free can be realized, which helps to improve the operation accuracy and stability of the laser auxiliary scanning mechanism.

Section 2 Analysis and Handling of Anomalies

I. Image Quality Issues and Solutions

During printing plate for export, quality problems may arise on the plate as follows:

(1) Computer: There are problems with the compatibility of the computer or there are viruses in the computer;

(2) Platemaking machine: Deviation of focal length and improper power; the laser is uncontrollable due to burrs during transmission or on the drum; screw track lacks oil and the mechanism is not adjusted properly;

(3) Processor: Development temperature and speed are set reasonably; the equipment cannot work properly due to improper maintenance and cleaning; the pressure of each roller shaft and brush is not adjusted properly; the matching ratio of replenishment fluid is unreasonable.

(4) Plate material: The plate material is not well sealed, the photoconductor is not evenly coated, the flatness of the plate material is beyond the standard, and the edges of the plate material are irregular, which is not suitable for the platemaking machine plate.

II. Other Abnormalities and Solutions

During operation, the hardware system should refer to the comprehensive debugging process and make adjustments according to the error report information.

1. Fault of Laser Unlock and Solutions

The reasons for Laser unlocking are as follows:

(1) The emitted light is not aligned with the center of the energy detection sensor probe;

(2) Dust exists on the surface of the lens, close-arranged, and energy detection sensor;

(3) Poor contact of the end of the wire;

(4) Laser drive plate failed;

(5) There is a failure in the constant temperature control of the laser box;

(6) There is a failure in the coupling of the fiber and the laser;

(7) Fiber damage;

(8) Laser damage.

2. Cheek whether the constant temperature control system of the laser box is in line with the process requirements

In the information prompt bar of program, there is a temperature prompt when locking the light and the temperature requirement is controlled in the range of 23.5~26.5℃；

The assembly requirements of constant temperature laser box are as follows:

(1) Open the software and execute the read power command. The temperature of the laser box is judged according to the feedback value, which requires a temperature of (25.5 ± 0.5)℃. In case of excessive temperature, it is required to check whether the cooling fan in the constant temperature box of the laser box is working properly, whether the constant temperature drive board is working properly, and whether the TEC is working properly.

(2) Check the working state of the cooling fan and observe the working state of the cooling fan in the constant temperature control box for the existence of a failed fan. The cooling fan requires replacement if it is present.

3. Make sure that the emitted light spot is located in the center of the photoelectric cell

(1) Reset the platform.

(2) Turn on a certain laser in the middle to observe the position of the light spot; the light spot is required to be located in the center of the light cell (visual inspection or light rotation of

Chapter 7 Test and Anomaly Handling of Platemaking Equipment

the position of the light cell; when the value of the laser power is maximum read by the software, that is the center position), as shown in Fig.7-1. In case of failure, the fixed position of the photocell needs to be adjusted to meet this requirement.

Fig.7-1 Test for emitted Light Point

(3) Apply the standard photovoltaic cell to lock the optical power. Replace the photocell with problems in case of failure.

4. laser power drop caused by Dust pollution

Dust pollution is the most common factor that causes a drop in laser power. The most pollution-prone positions of the UV model include the surface of the lens and the pollution of the light end of the close-arranged. Because the lens are used as the exposed part, it is very vulnerable to dust and other sundries. Therefore, it is a first necessary to check the laser lens for pollution when the laser power is dropping.

Before wiping the lens, the laser software is used to detect the laser energy value, and then the precision adjustment instruction is used to command the operation, so that the lens moves backward, and then the power is turned off, and the scanning platform is manually moved to the right limit.Use mirror-wiping paper or non-shedding cotton swabs dipped in a little detergent (10% diethyl ether and 90% anhydrous alcohol mixed) or directly use anhydrous alcohol to wipe along the lens in one direction The light spot of the photocell is also cleaned. After cleaning, turn on the machine to complete the precision adjustment instruction, so that the lens are restored, and then use the laser software to detect the laser energy value after cleaning. If the energy value is found to be lower than that before wiping, repeat the above cleaning work.

Fig.7-2 Wipe Lens

The close-arranged pollution can also cause the drop of laser power The cleaning process of close-arranged pollution is as follows:

(1) Use the command to move the cylinder seat to the reset point. As shown in Fig. 7-3.

Fig.7-3　cylinder seat

(2) Turn off the machine, remove the upper cover and the back upper cover;

(3) Disconnect the power cord and signal line of the plate unloading rack, loosen the fastening screws, and remove the plate unloading rack;

(4) Remove the cover of the scanning platform, and mark the line between the lens barrel and the close cylinder seat with a pencil. Ensure that the position does not change during recovery;

(5) Loosen the phillips screws on the protective cover of the lens barrel, and then rotate the lens barrel to remove it from the cylinder seat, as shown in Figure 7-4. Then clean the tight barrel surface, as shown in Figure 7-5;

Fig.7-4　cylinder separation

(6) Use non-woven fabric dipped in a little anhydrous ethanol to wipe the close-arranged head in one direction, Throw the non-woven fabric after use until the dirty points are totally gone.

(7) Power on, and then use laser software to detect the laser energy value to ensure that the energy value has increased Then the close-arranged head cleaning is done;

(8) Restore the lens barrel, reinstall the platform cover, reinstall the publishing shelf, and restore the machine cover.

Chapter 7 Test and Anomaly Handling of Platemaking Equipment

Fig.7-5 Cleaning the Close-arranged head

5. Laser power drop in case of a certain path or several paths

The failure of a certain path (assumed to be the X path) laser to lock may be due to the loose of the laser drive plate and head of the connecting wire, poor coupling of the laser to the fiber, damage to the fiber, and damage to the laser; it can be determined by using the interchange exclusion method.

(1) Determine the drive plate

Interchange the pin of the X path connecting to the laser with an adjacent path (lock the light correctly) and lock the light. If the power reduction follows the X circuit, it will be certain that the drive board card is not failing. Otherwise, the drive plate card should be checked, to see if the connection is reliable. Replace the drive board card if it still fails, as shown in Fig.7-6.

Fig.7-6 Checking Drive Plate Card

(2) Checking the coupling of the laser with the fiber

① Insert a straight screwdriver into the metal slot of the porcelain part and slowly drive the fiber to rotate at an angle, such as 60°;

② Test power: If the power backs to normal, then we can shake the fiber gently with a screwdriver. It is a coupling failure if the output power is stable and does not change.

(3) Check whether it is a failure of the fiber or the laser

① Fiber interchange can be used to determine whether it is a fiber failure or a laser failure, as shown in Fig.7-7 ;

② Gently push the ceramic core at the tail of the X path out of the laser with tweezers; operate the same way as the X ± N path (X ± N path is normal);

③ Clean the fiber head with a dedicated fiber head cleaner or with dust-free paper, on which no dust is visible on the diameter of the fiber;

④ Insert X and X ± N paths interchangeably;

⑤ The laser is locked, and the fiber close-arranged can be determined to be in trouble if the power of the X path is still low, then you need contact CRON Customer Service Department; If the X ± N path is low, it means the power follows the laser and the laser is malfunctioning and requires a laser change.

Fig.7-7 Checking of Fiber and Laser Failures

(4) Replacement of the laser

The anti-static device is required to protect the laser during operation.

① Pull down the connector to the laser; unscrew the pressure plate screw for fixing the laser with the inner hexagram handle, and then take out the laser and put it in place;

② Place the new laser in its original position. There is heat-conductive grease on the bottom of the laser. Then fix the laser with a pressure plate and tighten the nut with moderate torque.

③ Plug the laser into the corresponding position;

④ Clean the fiber head and observe it with a microscope; After being free of any rising points, interpose it into the inner hole of the laser;

⑤ Enable, test, lock the laser and make sure the power to meet the standard;

⑥ Rotate the fiber slightly without any change in power;

⑦ Operate according to steps ⑤ and ⑥ of (4), glue it for fixing, and pad it;

⑧ Reset the laser box;

⑨ Record the change.

(5) A method to determine the uncontrollable laser of a certain path

Run the procedure. In the "Power Compensation Configuration" window, double-click the value corresponding to 0 path. If it becomes to 50, it means that the laser power of the 0 paths is changed to 50% of the original lock light power. During imaging, the line corresponding to this

Chapter 7 Test and Anomaly Handling of Platemaking Equipment

laser will become thicker (positive plate) because of the lack of power in this path. The failed laser path can be identified by using this path as a reference. Click the ">" symbol button in the upper left corner to configure all percentage values together in the corresponding digital optical path. Lock the power and check that whether the lock is required.

Send the focal length test diagram to CTP. The head can be inspected by comparing the line part of the focal length test and putting it in advance. It can be compared from a staggering focus line if there is not a failure of a certain laser line in the 0 paths. The line of the 0 path is relatively thawed. If the 0 path is on the lower side, the adjacent on the left is the 2 path and the upper left is the 1 path. By analogy, find the laser that is uncontrolled. Find whether the connection terminals of this laser are loose. If the problem still cannot be solved, replace the laser drive plate. When observed with a microscope it seems to be an inverted image, which should be noted.

Section 3 Methods for Judging the Quality of Printing Plate Output

I. Referring to work instructions, process requirements

After receiving the printing process sheet, CTP publishers are required to adjust the document leading edge according to the requirements of the leading edge on the printing process sheet, after that, it can be published. Check that the leading edge is correct in the computer preview before publishing.

Before printing plate output, CTP publishers must carefully check the output list whether the "Product Name" to be output is consistent with the "Imposition Object Name" on the offset printing process sheet. After the test is consistent, sign the name of the examiner after the offset printing process sheet "Imposition Object" as the voucher for the test. Files that have not passed this test are not allowed to be output.

After the project output is set, place the printing plate and press the green button (or LOAD button). After the exposure of the printing plate is finished, it will be automatically entered into the processor, and the output printing plate will be placed on the shelf "To Be Tested" for inspection.

1. Self-inspection

After the printing plate is finished, the CTP output personnel will conduct a quality self-test on the printing plate to test the position of the plate, the direction of the leading edge, and the size of the leading edge; test the printing plate signal strip, dot restores normal, and the plate is free from stains; and put the printing plate on the "Qualified" shelf after the test is qualified.

2. Confirmation of CTP plate and other editions and special baking requirements

All products using CTP plates are not required to be spare printing plates for the onboard edition. The product using the CTP plate is larger than the 80000-seal should be baked.

3. Special inspection of CTP quality

The printing construction order or standard proof sheet shall be used as the inspection standard; the printing plate required for inspection on the same day will be inspected one by one by the inspector.

Check whether the color number of a set of printing plates is complete; the content of the "Target" at the bottom of the plate corner: including the number, product name, producer, specifications, color number, date, and whether the color code is complete. In particular, the "Product Name" must be in line with the "imposition Object Name" on the offset printing process. This inspection item is a necessary condition to judge whether the printing plate is qualified.

(1) Check whether the printing plate is in line with the specifications, color drawings, and text of the signed sample (standard proof printed);

(2) Check whether the whole set of printing plate's leading edge meets the offset printing process requirements on the printing construction sheet;

(3) Check whether each printing plate is correct in size and whether there are defects such as dirty points.

(4) Record it after the inspection is finished;

(5) Mark the product name, color and date with white chelate at the tow end after each printing plate passes the test;

(6) When a failed printing plate is detected, feedback will be made to the previous process for rework, repair, or reproduction. Do the identification and isolation work for the unqualified printing plate. A record should be made for the plate which needs to be discarded and destroyed;

(7) Printing plate distribution should be based on the order number, product name, color, vehicle number and other contents applied for in the receiving edition. The printing plate should be accurately distributed and recorded in the warehouse.

(8) Waste printing plates must be stacked centrally and disposed of uniformly per general waste disposal provisions;

(9) Printing plate produces the leading edge according to the printing process list;

(10) Storage conditions for printing plates should be controlled at temperature 17~26 °C, humidity 35%~70%, and the shelf life of the original packaging is one year.

II. Incorrect Imaging Focus Length

1. Initial Adjustment

Methods and Requirements:

(1) "Operate by Engineer"→"Modify Equipment Parameters" → Double click the value of

Chapter 7 Test and Anomaly Handling of Platemaking Equipment

parameter 19 (see Table 7-1), and the value corresponding to No.19 is changed to 1200 →"OK";

Table 7-1 Initial values of No.19 corresponding to thermosensitive / UV

Type of laser	The initial value corresponding to parameter 19
405 nm laser	1200
830 nm laser	1500

(2) Select "Template Adjustment" in "Command Function Operation", then select "Command Plate Load" to load a plate that is not exposed;

(3) Enter the LaserAdjust 4.10.exe program and turn on one of the lasers as normal-on;

(4) "Order Function Operation" →"Platform Movement" →"Send Command", move the recording head to the laser you can find the range of the drum with the plate (preparing a recording head such as a block to the range of the plate when moving, block the No.2204 sensor and then block the No.2203 sensor. Finally the recording head will stop at the specified position);

(5) Loosen the lock screw (two M3 hex socket screws) on the lens base. Move the lens back and forward and turn the drum slowly, observe the minimum laser point, and will cauterize a very fine white line on the sheet. At this time, it is the best focal length for the initial adjustment, and tighten the two screws;

(6) Check the fastness of all bolts on this structure;

(7) Turn off the laser and exit the LaserAdjust 4.10.exe program;

(8) "Command Function Operation"→"Command Plate Unload"→"Send Command".

2. Fine Adjustment of Focal Length and Object Distance

Enter the procedure and select the TIF file that is capable of a focal length test project, such as the provided Focal length test file "Test focus 745×605_2400_90", as shown in Fig.7-8.

Click "Operation"→"Open Output File" to the output project queue, right-click this project, select "Parameter Setting" in the drop-down menu, enter the project template setting interface, and check the "Focal Length Test", as shown in Fig.7-9.

Set the "Focus Test" (the recommended value is 20), whose unit "Time" represents the number of focus strips on the output plate and "Focus Compensation" (-40 is recommended).

Output focal length test diagram project according to the set parameters. Place head up, as shown in Fig.7-10.

A suitable focal length can be found on the plate by observing all the focal lines in the focal length test block evenly with no streaks, as shown in Fig.7-8. Or use a magnifying glass to go from right to left to observe and mark 1% ~ 2% of the dot from the beginning to the end; then count to the middle from both sides and take the middle one; then observe that the focal lines in the focal length test block should be all arranged uniformly. It is not normal to have a little thread or a little thicker. Dot of 1×1 (pixel×pixel) should be arranged evenly without obvious non-uniform depth; dot of 2×2 (pixel×pixel) should be absolutely uniform without any interference patterns, as shown in Fig.7-11.

Computer to Plate of Lithographic

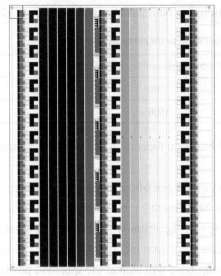

Fig.7-8 Focal Length Test

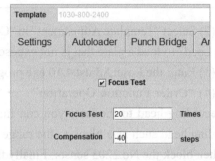

Fig. 7-9 Check "Focal Length Test"

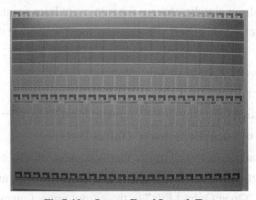

Fig.7-10 Output Focal Length Test

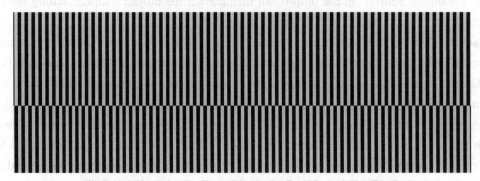

Fig.7-11 Observing the Focus Line in the Focal Length Test Block

For example, the plate is placed on the platform or desktop according to the tail side near the human body and is viewed from right to left with a magnifying glass. The 1% dot appears on the Mth; the 1% dot disappears after N; the region of N/2 should be the clearest in focal length. Then, the optimal position of the focus is M+N/2.

Note: M is numbered from 0. For example, if N is an odd-numbered value, you can see the lines in the small test blocks on the left and right of N/ 2. Verify by comparison that the corresponding change in focus length is (M+N/2)×4.

So the optimal value of the focal length is on the 12th block that has been numbered from the right side (counting from 0), then the change of the focal length is 12×5=60

$$f = F+(5 \cdot X) = -50+(5 \times 12) = 10$$

Observe the focal line with a magnifying glass (Fig.7-12). If there is a thicker line in the focal length line, then the object distance is too small, and the object distance parameter should be increased and otherwise reduced.

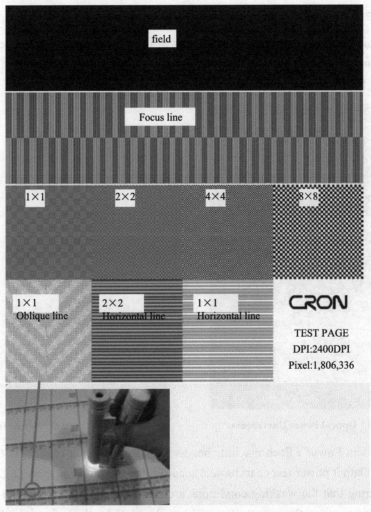

Fig.7-12   Focus Lines on Focal Length Test Block

Also, the edge of the field part should be observed. Adjustment is also required for the angle of the close-arranged (close-arranged and slanting-arranged way) in case of uneven edges. The process of adjusting the angle will affect the size of the object distance and focus length; in turn, the size of the object distance and focus length will also affect the change of the angle. Attention should be paid to this issue during adjustment.

There shall be no interference streaks in a flat net with an angle of 10%~99% at 90 °.

Enter "Operations by Engineer" →"Modification of Equipment Parameters" (Fig.7-13) and find the focal length value of the corresponding dpi for modification.

For example, it is a 2400dpi focal length test this time, the value of the focal length adjustment is 10, with the value of the original 19th parameter being 1200, and it will be changed to 1210 after adding. Click "Save Device Control Parameters".

Check whether tail creasing is sharp without avatars, and mainly verify whether the tail molding is compacted with the whole edge.

III. Optical Power Adjustment

Macroscopically, part 1×1 (pixel$\times$pixel) dot in Fig.7-13 shows whether there are streaks; See if there is any change in depth with a magnifying glass. If the answer is yes, compensation should be made for the optical power, as shown in Fig.7-14.

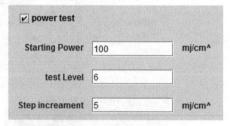

Fig.7-13   Optical Power Unevenness   Fig.7-14   Check "Power Test"

Set "Minimum Power". Each machine has its own proper power range; Set the appropriate starting power; Output power test chart project according to the set parameters (e.g.7-15).

After ensuring that the washing conditions are correct, pick the focal length bar of the dot with the best effect among the output plates (When the plate head up, the right dot is generally better at 1% of the dot, while the left dot is better at 99% of the dot. Select the dot focus strip with uniform tones). The up position of the head, count from the second focal length bar on the right to the left. Article X shows the best focal length. The current laser power is P, and the appropriate laser power after compensation of p 1 for each step is p, then the most appropriate laser power is

$$p = P + (p_1 \times X) \qquad (7\text{-}1)$$

Chapter 7 Test and Anomaly Handling of Platemaking Equipment

Measured by special instruments: the magnification of the dot is generally measured to be about 3 percentage points smaller than that in the theory; the bottom ash standard for blank areas is within 0.005 (a drop of acetone can be used in the gap, and an obvious color circle should not be found after spreading).

Cancel "Power Test" and select the laser power equal to the p-value in the drop-down box after the laser power; click "OK" to save this parameter, and change the template and the laser power to be output.

Fig.7-15 Instruments Test and Analysis

Section 4 Prepress Processing and the Relationship between Platemaking and Printing Quality

I. Quality Requirements of Photoconductors and Printing Plates

(1) Make sure that the film and CTP overprinting are accurate, with an error $\leqslant$ 0.05 mm; the field is flat and clean without obvious ink marks or miscellaneous grain; the dot is uniform and clear.

(2) It requires a tidy plate and color coding; no overlap of each color version of the signal strip, placed close to the bleeding, 5mm high; all colors are solid and fully spread. In the case of mixed competition, the color of the signal bar should correspond to the product.

(3) The cross-line and an angular line should appear within the scope of the paper, and should be appropriately lengthened if it cannot occur.

(4) The barcode is clear and complete and meets national standards. For the output of software film, please overlay the color to scan the barcode must be $\geqslant$ level B; for the output of CTP, the barcode part should first output the software film and implement it according to the software film testing standard.

(5) Each piece of assembly is in line with the sample requirements, i.e., the plate

arrangement and size meet the drawing or delineation requirements. Each pair shall do the trapping between spot color and spot color, spot color and four colors according to conventional requirements. For special requirements, please follow the notes or sample requirements.

(6) The target is required to be complete and positioned correctly; the top end of the standard is 25 mm from the net corner line.

(7) In the case of special leading-edge products, follow the comments or sample requirements.

II. Quality of Image Scanning

(1) Complete tone, clear calibration and high clarity.

(2) The scanned image should be loyal to or better than the original.

(3) The post-process should be considered in the graphic group.

(4) The output quality should be complete in font, accurate in picture link and correct in trapping.

(5) The prepress digital process version is as high and compatible with the latest version of PDF files.

(6) Digital proofing quality should be standardized on the printer. Select the correct characteristic file when printing.

Chapter 8　Checking Digital Proofs

　Objectives:

1. Be able to measure the technical parameters of the proof with a measuring instrument;
2. Be able to test the quality of the proofs with measurement and control strip ;
3. Be able to propose and implement weekly and monthly maintenance plans for prototypes.

Section 1　Proof Quality Inspection and Control Methods

In general, publishers or other customers only get proofs provided by printing plants. After confirmation, the printing plant officially starts printing the product required by the customer. In the process, the packaging designer, printing plant or proofing company bases film or electronic documents are provided by the printing company. The process of making the printed sample is called Proofing. The customer checks the plate design, and printing quality of the printed sample and signs to confirm that the sample can be used as the basis for printing the whole process is known as Sampling.

Digital proofing is a process that uses digital proofing equipment to reproduce printing colors based on the color range of the printing product and the same RIP data as the printing content and enables proofing to be made according to the actual printing status of the user. Theoretically, the color space conversion between the output devices can be carried out by software control when the digital proofing device color gamut is larger than the printing color gamut, so that the digital proofing effect simulates the color effect of the printing and output devices, i. e., the color management process of the digital printer imposition can help digital proof to replace the traditional proof signature. Digital proofing has evolved along with the development of CTP technology and is an indispensable part of a digital workflow.

Currently, the digital proofing system consists of a digital proofing output device and digital proofing control software. Among them, a digital proofing output device refers to any color printer that can be output digitally, such as color inkjet printers, laser printers, etc. However, most of the printers whose printing speed, space, screening method and product quality that can meet the publishing and printing requirements are large format color inkjet ones; digital proofing software includes EFI digital proofing software system, screen digital proofing system, Founder digital proofing system, etc.

I. the Control Process in the Digital Proofing

1. Device Correction- Printer Linearization

Ordinary color inkjet printers are all failures in linearity, which shows that for more than 90% of the dark colors and tone changes cannot be distinguished and then overlap appears. Moreover, the linearity of the primary color varies from printing to printing. If the property file washing appliance packaging of the proofread output device is made by the standard color table output from such a printer, it will cause an error in the equipment characteristics reflected by the property file of the output device. Therefore, printer linearization is the first step in implementing digital proofing color management. Note: The printer must be re-linearized after changing consumables such as paper and ink, or if adjustments are made artificially.

2. Property file production

The printer's characterization is an important part of color management. The basic process is to print a digital sample of a standard color code file with the standard color table file such as IT8.7/3 or ECI2002 through the digital proofing software and color printer. Tested and calculated by a photometer and special software, a property file that reflects the characteristics of a color printer and printed paper is generated.

3. Setup of Color Management in Workflow

After the printer is linearized and the property file creation of the printer is completed, it is necessary to set it in the digital proofing workflow to enable the digital proofing effect to simulate the color effect of the printing output device to match the digital proofs and printing proofs.

II. Factors Affecting Digital Proofs

Digital proofing refers to a new proofing technology that takes the digital publishing and printing system as the basis to process the page graphic information according to the printing and production standards and norms during the printing and production process and directly outputs color proofs. The ultimate goal of digital proofing is to simulate final printing effects, and the basic idea is to first establish a normal and easy-to-achieve printing effect as an ideal goal of digital proofing. Then adjust the digital proofing according to this effect to ensure that it is closer

Chapter 8　Checking Digital Proofs

to the ideal goal.

1. Printer

(1) Color gamut requirements

It is hoped that digital output devices will be capable of expressing printing effects in order to meet the needs of high-fidelity printing and personalized printing in the future. It is best to print in 8 or 9 colors for a wider color gamut performance.

(2) Printing accuracy requirements

The performance of the inkjet printer head will directly affect the output of the digital prototype. The printing accuracy achieved by the printing head determines the output accuracy of the digital proofing, so a low-resolution printer cannot meet the needs of digital proofing.

2. Ink

Printing ink plays a decisive role in restoring proofed color. The pigment ink is conducive to the preservation of the printing and the printing is not easy to fade away. The original color of its ink is closer to that of printing ink, but the influence of the light source environment on the color of the proof is more obvious.

3. Paper

The paper used in digital proofing is generally imitation paint printing paper, which has similar color expressiveness as printing paint paper, and is more likely to achieve the same color effect as printing; in addition, it is equipped with a coat suitable for printing ink, the quality of the coat will affect the performance of the proofs in terms of color and accuracy, while the ink absorption and rigidity of the proofing paper will also affect the quality of proofing.

III. Proof Quality Test Method

The methods of proof quality detection and control mainly include subjective visual detection and chromaticity detection.

1. Sujective Visual inspection

The proofs were observed by human eyes, and the quality of the proof was judged based on experience. It is greatly affected by human and external factors, which cannot fully reflect the quality characteristics of the proofs, but can judge the final quality of the proofs.

2. Chromaticity detection

Colorimetry is to get the color data from the test points on the proof picture directly. Calculating the chromatic aberration value of the corresponding points on the detection proof and the standard proof can help to judge whether the colors on the detection proof and the standard proof are consistent according to the size of the chromatic aberration value to judge whether the color is reproduced correctly at this position. Its advantages are that the judgment result of color is consistent with that of human eyes and of high accuracy, thus the influence of errors caused by subjective factors and fatigue when the human eyes judge color can be avoided. However, the change of the control factors during the printing process cannot be reflected when the chromatic aberration value is greater than the specified upper limit of chromatic aberration.

Section 2　Digital Proofing Measurement Control Strip

I. Fogra Digital Measurement Control Strip

A digital proofing authentication software is required to evaluate and certify the quality of digital proofing. Color management certification software uses data to confirm the correctness of color management for digital proofing. These softwares are developed by foreign companies to periodically certify the color management of digital proofing on data, which can improve the defects of digital proofing colors that are mainly verified by visual inspection. Unrecognized digital proofing systems can be adjusted in time to ensure color accuracy in digital proofing and thus color stability.

Digital proofing certification software generally selects international printing standards for certification, such as ISO12647-2, SWOP, GRACOL, and FOGRA. There are also corporate customized printing standards, and use the color measurement control strip defined by these standards to print them using digital proofing software. Generally, it can be printed on all four sides of the project.

In the color certification of digital proofing, 72 color blocks in total are used as the measurement control strip of FOGRA 39 V3 standard as the test strip of digital proofing software, as shown in Fig.8-1.

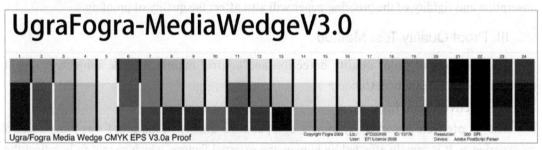

Fig.8-1　Measurement and Control Strip of FOGRA 39 V3 Standard

The digital printing measurement and control strip consists of three modules, Module 1 and Module 2, which are used to monitor the printing and reproduction process, and Module 3, which is used to monitor the exposure adjustment. The size of each measurement color blocks and control is about 6 mm×6 mm (sometimes it may be less than or greater than this value). Module 3 contains control blocks that are primarily used to monitor the exposure recording process of digital printing, so they are designed to correspond to PostScript measurement control strip, which is used by UGRA/FOGRA to test and control the output of the software film.

1. Module 1

It contains 8 solid blocks as follows: 1 cyan, 1 magenta, 1 yellow and 1 black; 3 "cyan + magenta", 3 "cyan + yellow", and 3 "magenta + yellow"; 1 "cyan + magenta + yellow". These color blocks are used to control the acceptable performance of the digital printing inks and the overlay printing effect of the three subtractive main colors.

2. Module 2

(1) Color balance control color block: This color block is the specified value of grey tone, which is related to the output of film. It contains two color blocks, of which the right color block is 80% black and is used to control the mesh tone screening effect while the left color block consists of 75% green, 62% magenta and 60% yellow, for comparison with 80% black color blocks. If the gray balance control is not ideal for printing, the color block will show color composition.

(2) Solid area: The solid area contains 4 solid color blocks which are arranged in the order of black, green, magenta and yellow. The color block is placed every 4.8 mm. The first solid color block (black block) is close to the color balance control color block, and its four corners, overprint is yellow which can help to check the printing color order, i. e., yellow ahead of black printing or black ahead of yellow printing.

(3) D control block: D is the Direction, so D control refers to direction control, which is to test the sensitivity of the combination of specific reproduction technology, reproduction equipment and printing materials in different directions screening.

The D control block is divided into four groups, cyan, magenta, yellow and black, and each group contains 3 color blocks with a total size of 6 mm×4 mm. When composing digital printing measurement control strip, it is usually arranged in the order of black, cyan, magenta and yellow. Located behind the solid color blocks. The 3 color blocks all apply line dot screening. The screening angles are 0,45 and 90 in order from left to right. The screen ruling is 48 lines per centimeter for each color block, with 60% (60% black). The reason for using a 60% order tuning value without intermediate tone value (50%) is that the color blocks after output are slightly darker than the intermediate tone, which enables a more clear identification of the direction sensitivity of the screening process.

Theoretically, the three color blocks should have the same density value when using the same screen ruling and dot shapes. If there is a large difference in the three density values measured: the combination of reproduction technology, reproduction equipment and printing material used by the user is too sensitive in a certain screening direction.

(4) 40% and 80% of the mesh control block: This control block also has 4 groups which include cyan, magenta, yellow and black. Each group of control blocks consists of 40% and 80% two color blocks with 150 lpi screening. The recording accuracy used by this number is consistent for most commercial printings. The two meshes regulate the distribution of color blocks to the intermediate dot in an asymmetric distribution, representing a slightly lighter (close to intermediate tone) than the intermediate dot and a near field percentage of tone. Different screening reproduction technologies for different digital printing processes will have different output effects. Therefore, these two control blocks can be used to evaluate the performance and behavioral characteristics of a specific digital printing screening technology and to measure whether the screening technology can achieve the desired recorded effect. When forming the measurement control strip combination, it is arranged in the order of black, cyan, magenta and yellow, positioned behind the D control block.

3. Module 3

The module consists of 15 gray blocks to varying degrees, each of which is the same size (6 mm × 10 mm) and is printed with black ink. The 15 color blocks consist of 5 rows, each containing 3 color blocks which apply a different dot structure. The ink coverage of the above color blocks is 25%, 50% and 75%, respectively, with 25% in the far-left column, and 50% in the second, third, fourth column and 75% in the fifth column.

The reason for printing these color blocks with only black ink is simple, which is to save the measurement control strip from taking up space on the page. The first line of the control block is always copied with the highest recorded resolution that can be achieved by the output device. The second line of the color block has a recorded resolution of half of the first line and the third line is one third of the first line. As a result, the larger dot structure can be found in the second and third rows of color blocks. The second, third and fourth column of the control block are all 50% black, and the second column is named 50 cb (Checker Board), and they are all plaid; The third column contains horizontal lines; and the fourth column contains vertical lines.

Theoretically, the order tuning value of each column of color blocks should be the same after module 3 is printed, with only the recorded resolution; in the line direction, the same order tuning value should also be found after 3 color blocks in the middle of each line are copied onto the paper. Therefore, if there is a difference in the tone of the 3 color blocks in the middle of each line, this difference must be related to the copy method, and the cause of the difference can be found from the perspective of the dot cable. The recording device should be adjusted to minimize the tone difference in the line direction at the output. When the color block's order tuning value in the column direction are different, it reflects the influence of screen ruling on the copy effect.

4. Module assembly

Module 3 of digital printing measurement control strip is used separately, and the assembly of module 1 and module 2 is free in principle. However, for a more regular arrangement, the following order can be used: balance control blocks can be arranged first. The next is the black, cyan, magenta and yellow plots, followed by the black D control block and the black tone screening control block. It is followed by black, cyan, magenta and yellow plots, followed by a cyan D control block and a cyan mesh dot screening control block. Then it is the black, green, magenta and yellow plots, followed by the magenta D control block and magenta mesh screening control block. Finally, there are black, cyan, magenta and yellow solid plots, followed by a yellow D control block and tone screening control block.

II. Digital Proofing International Standard ISO 12647-7:2016

The third revision of Digital Proofing International Standard ISO 12647-7:2016 was officially released on November 15, 2016. In December 2016, Fogra issued a notice reminding all Partners that the content of the new standard and the PSO certification is about to be updated to the new standard. In February 2017, CGS ORIS launched a new version of Certified Web V2.0.8, which supports this new standard. In April 2017, the Fogra Prepress Technology Department released Fogra Extra 36 after a period of testing, which explained the standard in writing to the industry. As we all know, ISO 12647-7 is the international standard of digital proofing, as shown in Fig.8-2.

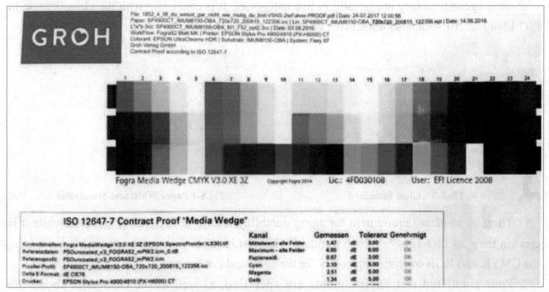

Fig.8-2 ISO 12647-7 International Standard Test Document for Digital Proofing

A copy of digital proofing, which is fully compliant with ISO 12647-7:2016, requires MediaWedge test color blocks to meet the tolerance requirements, as well as the corresponding requirements for the brightness, fluorescence content, wear resistance, and light resistance of the printing. The requirement that the luminance value L is greater than or equal to 95 for the three generic types of paper whiteness has become unideal shortly after the release of the 2013 ISO 12647-7:2016. Although it is more suitable for Fogra 39, it is not suitable for other proofing conditions, such as newsprints.

Therefore, the new ISO 12647-7:2016 standard puts forward the following requirements.

1. GLOSSY

First, the material is divided into three types: matte, semi-matte, and bright light. It should be the same type when the proofing material matches the printing material for color matching. The practice of using matte proofing paper as light powder paper printing following color drafts, or using glossy proofing paper as matte powder paper printing following color drafts is inappropriate, as shown in Figure 8-3.

2. Paper Simulation

The color of the proofing paper without graphic space should allow the simulation of the color of the printing paper (i.e., Absolute Colorimetric Intent), and the chromatic aberration $\triangle E_{2000}$ should be less than or equal to 3.0. To ensure the accuracy of the paper simulation, the luminance value (L value) of the proofing paper should be higher than that of the printing paper (L value), as shown in Fig.8-4.

3. Fluorescent Agent (OBA)

The fluorescence content level of the proofing paper should be the same as that of the printing paper. According to ISO 15397:2014, fluorescence agents are classified into four classes: micro, little, medium, and large quantities. There are cases that are free of OBA (fluorescence whitening agents), so they are usually in five categories. It is consistent with the classification in ISO 12647-2:2013 and also reflects the principles of digital analog printing.

Classification	75° ("TAPPI gloss")	60° (ISO 2813)
Glossy	> 60	> 20
Semimatte	20 - 60	5 - 20
Matte	< 20	< 5

Fig.8-3　Gloss Standard

Classification	Description of OBA
$0 \leq \Delta B \leq 1$	Free
$1 < \Delta B < 4$	Faint
$4 \leq \Delta B < 8$	Low
$8 \leq \Delta B < 14$	Moderate
$\Delta B \geq 14$	High

Fig.8-4　Paper Whiteness Simulation

There is no clear requirement for aging durability testing in ISO 12647-7:2013, while it is clear in the new ISO 12647-7:2016 standard. This test requires four same test proofs containing the CMYK and RGB overprinting colors for the solid and dot.

Four simulated test environments are specified, as well as the chromatic aberration requirements before and after processing in this environment. ① Room temperature environment:25 °C, 25% relative humidity, 24 hours;　② High temperature and wet environment:40 °C, 80% relative humidity, 24 hours; ③ Dry environment:40 °C, 10% relative humidity, one week; ④ Low Exposure Environment: At least 3 steps according to ISO 12040.

4. Chromatic Aberration Requirements before and after Processing

The paper whiteness chromatic aberration $\triangle E_{2000}$ is less than 2.5, and the maximum value of chromatic aberration $\triangle E_{2000}$ for other color blocks is less than 2.0. The chromatic aberration $\triangle E_{2000}$ can be relaxed to 4.0 in the case of matte light materials or other very rough materials.

The control color blocks chromatic aberration is an important change. The chromatic aberration formula adopts the $\triangle E_{2000}$, which is more consistent with the visual evaluation, rather than the $\triangle E_{76}$. The two formulas $\triangle E_{2000}$ (* 00) and $\triangle E_{76}$ (*ab) cannot be transformed from each other, so the new standard defines the new tolerance requirement.

The two chromatic aberration formulas are calculated differently. Therefore, digital contributions that meet the old standard may not meet the new standard, nor may they meet the requirements of the old standard.

Fig.8-5 is a comparison of parameter requirements of ISO 12647-7 new and old standards, including paper white simulation, all color blocks, tri-color gray, primary color (CMYK) $\triangle$ H, and primary color (CMYK) $\triangle$ E.

To evaluate the effectiveness and impact of the new and old standards, Fogra tested and analyzed 116 digital proofing, respectively. The results show that the data results of $\triangle E_{76}$ chromatic aberration formula are significantly different from the visual evaluation results for more saturated color blocks, while the new $\triangle E_{00}$ is more consistent with the visual evaluation.

Standard	Patch type	Tolerance
ISO 12647-7 old (2007)	Paper simulation	$\Delta E^*_{ab} \leq 3$
new (2016)		$\Delta E_{00} \leq 3.0$
ISO 12647-7 old (2007)	All patches	Maximum $\Delta E^*_{ab} \leq 6$ Average $\Delta E^*_{ab} \leq 3$
new (2016)		Maximum $\Delta E_{00} \leq 5.0$ Average $\Delta E_{00} \leq 2.5$
ISO 12647-7 old (2007)	Composed grey	Average $|\Delta H| \leq 1.5$
new (2016)		Maximum $\Delta C_h \leq 3.5$ Average $\Delta C_h \leq 2.0$
ISO 12647-7 old (2007)	Primary colours	Average $|\Delta H| \leq 2.5$
new (2016)		Maximum CMY $|\Delta H| \leq 2.5$
ISO 12647-7 old (2007)	Primary colours	Maximum $\Delta E^*_{ab} \leq 5$
new (2016)		Maximum $\Delta E_{00} \leq 3.0$

Fig.8-5 Comparison of Parameter Requirements in ISO 12647-7 New and Old Standards

5. Spot Color Evaluation (CxF)

This is the first time that the proofing process standard defines the requirements of the spot color evaluation. Spot color is very common in packaging applications. Especially in response to high customer's expectations of color consistency, it is often only the spot color that can be adapted to meet customer's requirements.

However, the current digital proofing system generally uses an inkjet prototype and does not allow the addition of custom spot color ink, color gamut is always limited. Therefore, this is only applicable to cases where the spot color to be evaluated is within the range of color gamut that can be presented by the prototype.

In this case, the new standard of ISO 12647-7:2016 specifies a maximum of chromatic aberration $\triangle E_{00}$ for reference color samples and proofs within 2.5, while the reference color samples are selected or provided by the customer, and the standard recommends the use of CxF/X-4 data.

The spot color definition based on CxF/X-4 is more desirable. It should contain the spectral data of spot color, the data of black/white, the transparency of the ink, etc. All of these can improve color communication between customers and printing mills. It is still too early to standardize the spot color. Therefore, it is highly recommended to negotiate with the customer beforehand, especially in the case where the spot color is outside the proofing color gamut. It is more feasible to define the Lab value of spot color, but this approach is limited to 100% field. The definition of the dot can only be based on CxF/X-4 data.

6. Other Requirements

The CMYK color blocks representing the printing status should be printed in each prototype. The standard specifies the color block attributes that should be included, such as four colors and overprinting colors on the field, dot, gray balance color blocks, etc., which are defined in ISO 12642-2.

Typically, it is not required to make control strips, but it is required to be clear about the target standards to be matched and to choose the corresponding internationally renowned printing agencies that have existing control strips, such as IDEAlliance ISO 12647-7 Control Wedge 2013 and Ugra/Fogra Media Wedge CMYK V3, etc. (these two are the new version of the 3-line control strips, and it is not recommended to use the old version of the 2-line control strip).

A digital sample is not standard if there is only a picture and text without additional information, which will often bring great confusion and trouble to users. The ISO 12647-7 standard specifies in detail that proofing should include the following information:

Tagging information, execution standard version ISO 12647-7: XXXX; file name, digital proofing system name, type of printing material, simulated printing conditions, time and date of proofing, measurement conditions, M0, M1 or M2, type of colorant, color management ICC Profile used, RIP name and version used, scaling ratio, type of surface processing, date and time of latest calibration, details of any data preparation, application noise information, etc.

The digital proofing system still suffers color shifts although it is relatively stable. ISO 12647-7 standard specifies that the CMYK field and 50% dot, as well as RGB, are required to be re-evaluated with a maximum chromatic aberration $\triangle E_{00}$ of not more than 2.0. The same instrument should be selected to measure the same position, and the proofing system should be recalibrated if necessary.

In addition to the above requirements, there are also wear resistance tests, gloss requirements, dot requirements, no streaks, image overprinting and resolution, etc.

In short, standard digital proofing product needs to simulate the actual print colors and visual appearance as much as possible, as well as sufficient proof information for the convenience of the user. Attention should be paid to whether the user's requirements can be met in the selection of consumable materials, instrument configuration, measurement mode, software setting, color correction, and standard selection. In particular, attention should be paid to whether all aspects of the proofing system can meet the requirements of the new standard when translating into new standards such as Fogra 51/52, GRACoL 2013, etc. It is with qualified and accurate proofing that printing can be better guided and the roles of digital proofing can be maximized.

Section 3 Method of Application of Measurement Instruments

I. Densimeter

Density measurement and densitometer, as the most important special instrument in the quality evaluation of color copies, or as a means of inspection and control during the reproduction process; are inseparable from the evaluation of process technology and raw materials, It is closely connected with identification and evaluation of color quality during production.

The principle of densitometric measurement is close to that of visual identification by printing workers. Fig.8-6 is a schematic of the working principle of densitometric measurement. Light emitted from stable light source 1 is focused by lens 2 on the printing surface, with a part of the light absorbed. The amount of absorption depends on the thickness of the ink film 5 and the concentration of the pigment, and the unabsorbed light is reflected by the printing paper. The lens system 6 collects and measures reflected light of 45° from the light and transmits it to the receiver (photodiode) 8. The photodiode converts the received light into power. Electronic system 9 compares this measured current with the reference value (the reflected amount of "standard white"). The absorption characteristic of the ink film measured is calculated according to this difference. The results of the ink film measurement are shown on the display 10. The color filter 4 on the light path only allows light with the corresponding wavelength of printing ink to pass.

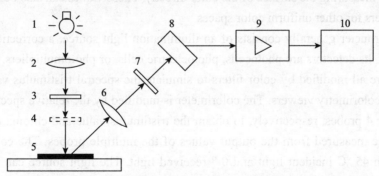

Fig.8-6 Structure and Measurement Working Principle Diagram

The densities measured by three filters, red, green and blue, are called color densities or three filter densities, which are expressed by D_R, D_G, and D_B respectively. D_R reflects the absorption degree of the colorant to the red light in the incident light spectrum. Similarly, D_G and D_B indicate the degree of absorption of the pigment to green and blue light in the incident light spectrum respectively. Therefore, the three independent parameters of color density, D_R, D_G, and D_B can accurately represent the color attributes of a color swatch.

II. Spectrophotometer

A spectrophotometer is used to measure the reflectivity of the color surface to each wavelength of light in the visible spectrum. The light of the visible spectrum is illuminated on the color surface at a certain step distance (5 nm, 10 nm, 20 nm), and then the reflectivity is measured point by point. The relationship between the reflectivity values of each wavelength of light and each wavelength is plotted to obtain the spectral photometry curve of the color surface to be tested. The measured values can also be transformed into other superficial color system values, each of which uniquely represents a color.

The spectral photometer consists of a light source (usually a halogens tungsten lamp or xenon flash lamp, i.e., standard light source A), a single-color device (prism or crown), a receiver (photomultiplier tube), and a recording instrument (potentiometer). The principle is to decompose the light of the light source into a spectrum and use a slit baffle to guide a single-color band from the obtained spectrum and then project the single-color band onto the sample to be tested. The transmission or reflection of mono-colored light through the object to be measured is shown on the record table. For the color to be measured, the transmission coefficients of light reflection coefficient and absorption coefficient are generally measured from the range of 400 ~ 700 nm of the visible spectrum every 10 nm. The reflectance or transmittance of each wavelength is connected with the points to draw the spectral photometry curve. The spectral photometry curve can represent the complete color characteristics of a colored object, which has only one spectral curve.

III. Colorimeter

A colorimeter enables the users to obtain a visual response proportional to the three color excitation values X, Y, and Z, and obtain the X, Y, and Z values of the measured color through conversion by measuring the surface of the color directly. These values can also be translated into color parameters for other uniform color spaces.

The colorimeter generally consists of an illumination light source, a correction color filter, and a detector. Its detectors are photocells, photoelectric cells, or photomultipliers. Their spectral sensitivities are all modified by color filters to simulate the spectral tristimulus value curves of CIE standard colorimetry viewers. The colorimeter is modified by the relative spectral sensitivity curves of 3 or 4 probes, respectively, to obtain the tristimulus value and chroma coordinates of the color to be measured from the output values of the multiple probes. The colorimeter was measured with 45 °C incident light and 0 ° received light. The light source can be a halogens tungsten lamp or thawing flash lamp, standard light source A.

The colorimeter is a density meter with three broadband color filters. The absolute accuracy of the color measurement value is not ideal due to certain errors in the instrument's own devices and principles. However, it is still a widely used color measurement instrument due to its price advantage.

IV. Use of Color Spectrophotometer

1. Connection to the Spectrophotometer

Connect the power wire to the power jack of the spectral densitometer. Turn on the power switch on the instrument and keep the display of the instrument highlighted.

2. Calibration of the Spectrophotometer

Place it under the measuring head of the measuring instrument when using the instrumental whiteboard. It should be noted that a whiteboard for correction is required for each instrument. It is required to make sure that whether the serial number of the whiteboard used matches that of the measuring instrument before calibration. Also, make sure that the surface of the whiteboard is not contaminated, and that the testing area of the measuring instrument corresponds to the center of the whiteboard.

Then, select the "Calibration" function of the measuring instrument through the menu and press the measuring head to wait for the successful calibration.

3. Density Measurement

Select the "Density Measurement" function (press the "Enter" arrow) by selecting the (up/down) "Move" arrow, on the instruments panel. There are differences in relative and absolute densities when measuring density. Relative density refers to the density value obtained by subtracting the measured density from the paper density when measuring density; absolute density refers to the density value obtained by measuring the sample directly when measuring density. The choice of two different types of density measurements can be achieved by adjusting the mode in the "Option" on the measuring instrument.

Also, when it is necessary to compare the density difference between the sample and a standard sample, the density value of the standard sample can be measured and defined as a reference. The density difference between the sample and the standard sample can then be obtained by selecting the way of subtraction of the density value and the reference value (DEN—REF0X) during the measurement. This function uses the reference item in the Option to select the series number of the relevant reference standard used and changes the mode of operation of the "Density Measurement" function to density—reference (DEN—REF0X).

4. Other Measurement Functions

In addition to the measurement of color sample density, the spectrodensitometer can also measure important indicators related to printing quality evaluation, such as dot area, printing contrast, overprinting and hue error and gray-scale. Also, the color values of color samples can be measured by some types of photometers, such as CIEXYZ and CIELab and chromatic aberration. For other measurement functions, please refer to the website of the relevant brand and model.

V. Use of Color Spectrophotometer

For example, the X-Rite Eyeone instrument is used to introduce the use of the color spectrum apparatus.

1. Instrument Connection and Calibration

Connection is made between the power line and the power jack of the spectro photometer, and the data line is connected to the data communication port of the computer (the new spectro photometer mostly uses a USB interface to connect to the computer).

Turn on the power switch on the instrument and keep the status LED of the instrument is on. Place it under the measuring head of the measuring instrument or in a specific calibration area by/via the instrument's calibration white board. As with the densitometer, each instrument corresponds to a whiteboard; It is required to make sure whether the serial number of the whiteboard used is consistent with that of the measuring instrument before calibration; ensure that the surface of the whiteboard is not contaminated and that the testing area of the measuring instrument corresponds to the center of the whiteboard. Calibration of the instrument is done through the prompts from the application.

2. Measurement of Spectrophotometric Values

The spectrophotometer works in a variety of ways, such as manual measurement or automatic scanning measurement.

When working, the model of the instrument should be selected first according to the instructions of the measuring software via the measuring software. If the instrument model selected is the same as the instrument model connected to it, the status will display "OK", which means the measured data is the spectral value or chroma value of color (the measured value is the spectral value, which means Spectral is checked in the instrument connection window, otherwise, the measured result is the chroma value).

Then, the color table (Test Target) required for the measurement needs to be defined in the measurement software. This color table enables you to select the common standard color table that has been defined in the measuring software, as well as to define the number of colors measured via custom, and the number of times each color block is sampled and measured. The measuring software will then prompt for the instrument to be aligned with the calibration board.

After the calibration is completed, the instrument will calculate and determine the position of moving measuring head each time according to the positioning result of the color table and read measurement result when the measurement is carried out. The manual measurement requires a manual aiming the measuring head at the color blocks. Press the Measure section to complete the measurement. As shown in Fig.8-7, the working interface of MeasureTool driving spectral photometry in the measurement software ProfileMaker.

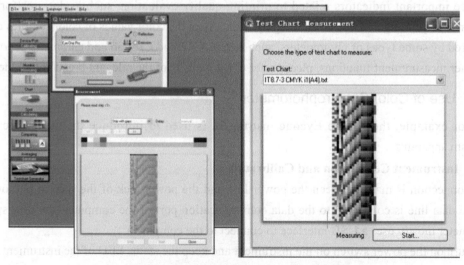

Fig.8-7　MeasureTool Measurement Interface

Chapter 9 Lithographic Printing Standards and Management

 Objectives:

1. Be able to do the grade evaluation of the print;
2. Be able to apply knowledge of the quality management system to achieve quality statistics, analysis and control during operation; competent to propose corresponding plans for problems that may arise during proofing;
3. Be able Competent to make the quality management plan of the sampling process according to ISO-9001 standards;
4. Be Competent in production planning, scheduling, equipment safety and personnel management;
5. Be able to develop environmental protection project measures of the department; Able to develop and optimize the process of platemaking;
6. Be Competent in the development of special process plans;
7. Be able to formulate production plans according to each process;
8. Be able to analyze the reasons for the quality issues.

Section 1 Standards of Related Printing Quality

I. ISO 12647 Standard

ISO 12647 Printing Process Control Standard is developed by TC130, a special committee on printing technology in the International Standardization Organization (ISO). The international standards it sets can be grouped into the following categories: Term Standards; Prepress Data

Exchange Format Standards; Process Control Standards for Printing; Adaptation Standards for Printing Raw and Auxiliary Materials; and Human Engineering/Safety Standards. Among them, the Printing Process Control Standard is a basic standard widely used in printing and production, which occupies an important position in ISO/TC130. It specifies the technical requirements and testing methods for key quality parameters in the printing process. For example, a series of standards of ISO 12647 are grouped according to the process methods of offset printing, concave printing, screen printing, soft printing and digital printing. Their respective technical requirements and inspection methods are stipulated for the quality control parameter of different printing methods.

The ISO 12647 standard (process control of printing technology, mesh adjustment color separation, proofs and finished printed products) was developed based on comprehensive data from several national printing quality committees around the world. It is a series of standards of packaging anti-counterfeiting, which provide the smallest set of parameters for technical attributes and visual characteristics produced by various printing processes (offset printing, gravure printing, flexible printing and so on). It provides guidelines for manufacturers and printing practitioners to help set the equipment to a standard state. Measured data from these printing factories can be used to create a ICC color description file and generate proofs that match the printed color.

ISO 12647 is a standard based on standards that include ink, paper, measurement and visual observation conditions. This standard covers many parts. In each part of the ISO 12647 standard, the minimum value of this process parameter is defined for a different printing process.

ISO 12647-1: Process control for the production of mesh tone color separation proof and printed product: parameters and measurement methods, packaging decoration

ISO 12647-2: Offset lithography

ISO 12647-3: Coldset offset lithography on newsprint

ISO 12647-4: Publication gravure printing

ISO 12647-5: Screen printing

ISO 12647-6: Flexographic printing

ISO 12647-7: Sample producing, digital printing and proofing directly based on digital date

In short, ISO international standards can facilitate the exchange of information among printing practitioners, prepress, and print buyers. As an international standard, it has been increasingly accepted and adopted by printing buyers and has become a pass for the import and export business of printing enterprises.

In China, there are more and more printing enterprises that adopt international standards, such as Beijing Shengcaihong Platemaking Printing Technology Co., Ltd., Zhejiang Yingtian Printing Industry Co., Ltd., etc., and even some enterprises that are not yet very large, all of which execute each step of the printing process in strict accordance with a series of international standards such as ISO 12647-2, ISO 12647-7, ISO 2846, ISO 12646, ISO 10128, etc., so that products can be followed and checked in an orderly manner to ensure product stability.

Chapter 9 Lithographic Printing Standards and Management

II. ISO 12647-2 Standard

ISO 12647-2 belongs to/refers to the standard "Graphic technology — Process control for the production of halftone color separations, proof and production prints — Part 2: Offset press process control" in ISO 12647, whose credibility of ISO 12647-2 is based on the solid density block and TVI (dot gain curve) curve.

1. Solid CMYK Chroma Standard

Printing quality control analysis reveals that printing density indicators fail to accurately control the quality of printing four-color inks along with the development of printing and testing technology. Therefore, ISO 12647-2 Printed process control standard has revised the evaluation index of printing four-color field color to the value based on CIELab color value in a new standard issued in 2004, as shown in Table 9-1.

Table 9-1 ISO Printing Solid Standard

Paper type	1+2			3			4			5		
	L*	a*	b*	L*	a*	b*	L*	a*	b*	L*	a*	b*
On Black Backing												
Black	16	0	0	20	0	0	31	1	1	31	1	2
Cyan	54	-36	-49	55	-36	-44	58	-25	-43	59	-27	-36
Magenta	46	72	-5	46	70	-3	54	58	-2	52	57	2
Yellow	88	-6	90	84	-5	88	86	-4	75	86	-3	77
Red (M+Y)	47	66	50	45	65	46	52	55	30	51	55	34
Green (B+Y)	49	-66	33	48	-64	31	52	-46	16	49	-44	16
Blue (C+M)	20	25	-48	21	22	-46	36	12	-32	33	12	-29
On White Backing												
Black	16	0	0	20	0	0	31	1	1	31	1	3
Cyan	55	-37	-50	58	-38	-44	60	-26	-44	60	-28	-36
Magenta	48	74	-3	49	75	0	56	61	-1	54	60	4
Yellow	91	-5	93	89	-4	94	89	-4	78	89	-3	81
Red (M+Y)	49	69	52	49	70	51	54	58	32	53	58	37
Green (C+Y)	50	-68	33	51	-67	33	53	-47	17	50	-46	17
Blue (C+M)	20	25	-49	22	23	-47	37	13	-33	34	12	-29

2. TVI Standard

The printing dot gain is described in ISO 12647-2 Offset Printing Control Standard. The dot gain values TVIs at 50% tone for ISO standard paper and offset printing conditions are shown in Table 9-2. The standard dot gain curve obtained by ISO according to the specific printing conditions is shown in Fig.9-1.

Table 9-2 Dot Gain Values at 50% Tone for ISO Standard Paper and Offset Printing Conditions

Printing conditions	TVI (for different screen ruling)		
	52LPcm	60LPcm	70LPcm
Colorful printing dot gain in four colors [1]			
Positive plate [2], paper type [3] 1, 2	17	20	22
Positive plate, paper type 4	22	26	—
Negative plate, paper types 1, 2	22	26	29
Negative plate, paper type 4	28	30	—
Four-color plate dot gain in rotation commercial printing and other printing conditions [1]			
Positive plate, paper types 1, 2	12	14 (A) [4]	16
Positive plate, paper type 3	15	17 (B)	19
Positive plate, paper types 4, 5	18	20 (C)	22 (D)
Negative plate, paper types 1, 2	18	20 (C)	22 (D)
Negative plate, paper type 3	20 (C)	22 (D)	24
Negative plate, paper types 4, 5	22 (D)	25 (E)	28 (F)

Note: ① The black plate has the same increase value or greater than 3% as other color plates.
② Types of printing plates should be independent compared to direct-to-plate technology, but different control parameters are often used to output positive and negative printing plates in the actual production process.
③ The type of paper is specified in the ISO 12647-2 offset printing standard.
④ A, B, C, D, E, and F are the standard curves corresponding to Fig.9-1. The curve is calculated by measuring the chroma values of the CMYK color samples on the printed copy to obtain the dot value. Then draw a curve.

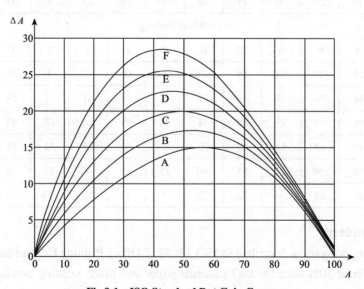

Fig.9-1 ISO Standard Dot Gain Curve

Section 2 Configuration Standards of Lithographic Platemaking Equipment

I. CTP Network Environment

Memory: 2G or larger;

Hard Disk: Mechanical hard disk 7200r or above, hard disk size is 1T or above. Use SSPS for software running disks (including TIFF files for import) and system disks.

Graphics card: 256M or larger;

Operating System: Windows XP/Win7/Win8/Win10;

USB: 2.0 Port;

CPU: Intel dual-core E5200/7400 or above.

II. Environmental Conditions Acceptance

The operating temperature of the space in which platemaking machine is located must be between 20 ~ 30 ℃. Humidity:40%~60%. The equipment site is equipped with thermometers and hygrometers.

The ground flatness of the equipment site must be within ± 4 mm.

The special ground wire should be prepared for the equipment, which requires that the voltage must be stable. The diameter of the ground line is 4 mm^2 (the diameter of zincing metal should be not less than 10 mm, and the deep-buried wet soil should be less than 1.5 meters), and the resistance of the equipment to the ground is $\leqslant$ 0.5 Ω. Also, the computer case should be grounded as well. Prepare a 4 mm^2 grounding wire of a certain length to ground the equipment and the computer shell at the same time.

The equipment is equipped with a UPS-continuous power supply (> 6 kVa).

The external connection of the equipment is equipped with a 35A air switch and a standard power cable.

There shall be no large equipment around the equipment that will affect platemaking machine due to vibration.

The dust removal measures and meeting requirements of the workshop should meet the national quality secondary standard, i.e., the API value should be > 50 and $\leqslant$ 100.

A water purification device is required for poor water quality in the equipment site. In case of insufficient water pressure, a water pump is required.

III. Equipment Transportation Inspection

After the equipment goes to the customer, it is first checked to see if there is damage to the appearance of the wooden box, collision and whether the overall wooden box is complete;

After dismantling the wooden box, check whether the surface of the equipment has paint shedding, scratches, concave shells and watermarks, or whether there is a shift of the upper lid, etc.

Check whether the supporting software, toolkits, parts kits, simple plate set, etc. are complete according to the random list.

IV. Installation & Debugging Inspection

Platemaking machine level test: Three lifting corners of platemaking machine are raised to ensure that the ground wheels are relaxed and a horizontal ruler is placed on the surface of the drum to confirm the left and right level; then the horizontal ruler is placed on the surface of the wall to confirm the front and back level.

Processor level test: It can be horizontally adjusted according to the liquid in the developing tank.

Confirm all fixing ties and removal of fixing devices.

Verify that all movable components are normal, such as the balance block, plate pressing roller, etc.

Processor brush pressure test: Cut a pair of PS plates into the lower part of the brush and adjust to the pressure suitable for contact; the pressure is consistent between the left and right.

Processor rubber roller pressure test: When the pressure adjustment screw just came into contact with the rubber roller sleeve, both sides were slightly rotated until about 90 degrees on both sides.

Processor function test: Confirm with a thermometer that the actual temperature of the solution is consistent with the display temperature; observe whether the circulation state of the solution is normal; add refrigerant, add solution set the temperature and washing time according to the proportion of the supporting materials for the solution, and the concentration of replenishment fluid should be higher than the developing liquid ratio of developing tank.

V. Platemaking Machine Function Test

(1) Confirm the normal connection of the UPS's uninterruptible power supply.

(2) Equipment self-test: The engineer should closely observe whether the relevant equipment has abnormal movements and abnormal sounds when the equipment carries out corresponding actions.

(3) Verify that all parameters are at their best value.

(4) Laser testing: Verify that the laser power meets the factory standard. Confirm that laser temperature is within the normal range ($25°C \pm 3°C$).

(5) The exposure output confirms that the air pressure of the drum is in line with the range value.

(6) Printing plate test: Confirm that the range value error of the dot is within 1%; confirm that 1%~99% dot is restored; confirm that the blank part of the printing plate is free of bottom ash; confirm that the image leading edge conforms to requirement; confirm that the image is centered and that the surface of the printing plate is free from any scratches; confirm that the error of four colors overprinting is within 0.01 mm.

Chapter 9 Lithographic Printing Standards and Management

VI. Causes and Solutions of the Quality Issues in Printing Plate

1. Insufficient or excessive exposure capacity causes dot enlargement or loss

Solution: Adopt energy testing to choose the best energy.

2. Bottom ash or dot loss due to insufficient washing or over-washing

Solution: adjust the washing time, supplement the developing liquid, and change the liquid regularly.

3. There are scratches on the printing plate

Solution: It may be that the printing plate has its own wounds and the printing plate will be changed; it may be caused by the processor, which is usually a crystalline scratch on the surface, or aging and deforming of the rubber roller; cleaning the rubber roller or changing the deformed rubber roller.

4. Leading edge is in the wrong position or the image is not centered

Solution: Calibrate leading edge parameters and image parameters.

5. Image dislocation

Solution: Typically, external interference is caused and the grounding is verified to be in good order, with the power and data wires bundled separately.

6. Dot virtue non-energy and liquid problems

Solution: The surface of the lens or photovoltaic is contaminated, cleaned or the focus parameters deviate, and the focus parameters are calibrated.

7. Inaccurate four-color sleeve

Solution: Adjust the accuracy of side gague duplication positioning to achieve 0.01 mm positioning accuracy.

8. White lines appear in the field part of the image or black lines appear in the blank part (refers to the color of the drug film)

Solution: The laser control panel may be replaced when a certain path of the laser is not under control.

9. Part of the printing plate exposure and then stop the exposure

Solution: It may be a data transmission failure. Replace the USB interface or replace the USB cable.

Section 3 Process Control

G7 Printing Certification is made by the U.S. General Requirements for Applications of Commercial Offset Lithography (GRACoL), in combination with years of practice in CTP equipment, to explore and conclude. It aims to achieve a consistent effect of commercial offset printing quality with the assistance of CTP.

I. Preparation

A total of two printing operations are required for the estimated length of time and the whole process of work, and they are correction base printing and characteristic printing. It will take about one to two hours each. Half an hour to an hour is required for printing plate correction and one or half a working day is needed in total. All the work should be arranged on the same day and be done by the same operators on the same equipment and materials.

1. Equipment

The printing press is required to be debugged to the best working condition, consumables as well. Check for compliance with the related objectification parameters. The focus length, exposure and chemical of the CTP are adjusted as required by the manufacturers. Publish with a natural curve without linearization correction.

2. Paper

Use ISO1 # paper without fluorescence as much as possible. It requires about 6000 to 10000 pieces of paper, depending on the efficiency of the operation.

3. Ink

Use ISO2846-1 ink.

4. Standard proofs

The preset *GRACoL7 Printing Press Calibration Model* can be purchased from the www.printtools.org website or can be done by oneself, as shown in Fig.9-2.

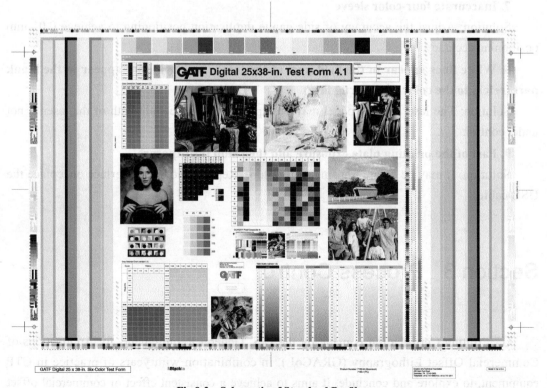

Fig.9-2　Schematic of Standard Profiles

Chapter 9 Lithographic Printing Standards and Management

The standard proofs should include:

(1) Two copies of P2P23 ×standard (or later versions), forming 180 degrees from each other;

(2) GrayFinder 20 standard (or a later version).

(3) Two copies of IT8.7/4 characteristic standard samples, mutually 180°, in a row;

(4) A signal bar with a length of half an inch (1 cm) (50C, 40M, 40Y) of full paper laid horizontally;

(5) A signal bar with a length of 50K and a half an inch, (1 cm), of full paper laid horizontally;

(6) A suitable printing press control strip, which should include some important parameters of G7, such as HR, SC, HC, etc.;

(7) Some typical CMYK images.

5. Others

Other equipments include NPDC drawings provided by GRACoL for free (as shown in Fig.9-3); printing plate dot meter for measuring printing prints; spectral photometer; D50 standard observation light source; curve tools for drawing, etc. It is also possible to purchase the software IDEAlink from GRACoL to facilitate quick completion of the testing.

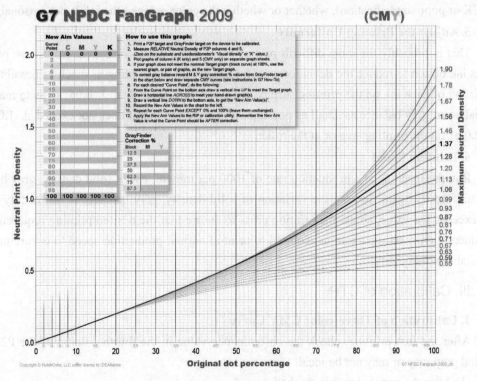

Fig.9-3 Schematic of NPDC Drawing

II. Calibration of Basic Printing

1. Printing Conditions

The printing press and its consumable materials should be adjusted correctly, including the

163

viscosity of the ink, rubber blanket, cylinder-packing, pressure, fountain solution, environmental temperature, humidity, etc. The suggested color order for printing is K-C-M-Y. It is best not to use a drying system of the machine.

2. Solid Density (SID)

Print according to the color value (L*a*b*) or density value of the standard field ink (data shown in Table 9-1).

3. Dot Gain Carve

TVIs were measured for each CMYK color plate. The difference between each Dot Gain Carve of CMY should be within ± 3% and slightly higher than 3%-6% in black.

4. Grey Balance

With a spectro photometer set to D50/2 °, measure the gray balance values of several HR (50C, 40M, 40Y) blocks on the printing sheet. The ink solid density of the CMY is adjusted to show the error combination of L*a*b*.

If the gray balance cannot be close to the target a*, b* value, or cannot be adjusted by a small amount of solid density, then check whether one or more TVIs are too large, whether the color of the ink is not proper, the overprinting is not ideal (which may be due to ink stickiness TACK or poor emulsification), whether or whether the color order of the ink is not reasonable.

5. Adjust the Printing Uniformity

This is probably the most difficult part of the printing press correction. Adjust the printing press ink button to reduce the deviation of solid density on the printing as much as possible. The deviation of each ink on the printing surface shouldn't be greater than ± 0.05, so as to make the deviation of gray balance as small as possible. It is preferred that it does not exceed ± 1.0 a* or ± 2.0 b* on the printing area or at the roller.

6. Printing Speed

Run the machine at a production rate of 1000 per minute; check solid density, gray balance and uniformity again. In case the variation of solid density, gray balance, or uniformity of the ink exceeds the value, adjust the printing press to ensure that the desired printing requirement is obtained; then speed up again as required to print at normal production speed to ensure uniform and stable print quality.

III. Calibration of CTP

1. Calibration of Three-color CMY Curve

After the first printing, check the chrominance value in the fourth column of the P2P. The neutral gray may or may not be ideal.

(1) After the gray balance is reached

① Measure P2P: Select the print that meets the requirement, and then measure the numerical relative density value in the fourth column of P2P after drying. At least two readings should be measured from different regions to take their averages.

② Draw the actual NPDC curve: Download the free drawings on the official website of

GRACoL to make the actual NPDC curve plot for printing and output.

③ Determine the standard curve: Locate the target diagram that is closest to the solid density value of actual production in the NPDC sector graph. If not, you can draw it by yourself with curve plate analysis from the two close curves at the top and bottom.

④ Determine correction points: Check the actual curve to see where the curve is the sharpest, and then determine the curve points to be corrected. As the human eye is more sensitive to the bright tone, it is best to set up more points at the bright tone.

⑤ Correct NPDC curves: Draw a vertical line up at each correction point from (60,44), intersect the target line and draw another horizontal line from the intersection point, and intersect with the standard line (to the left or the right) to draw a vertical line down from the intersection and intersect with the coordinate to obtain a new target value. Record this value on the drawing and repeat the above steps at each curve point. Keep it at 0% and 100%.

(2) Lack of gray balance

① GrayFinder: Use GrayFinder to complete the correction if the desired gray balance cannot be achieved during printing. The middle of the color blocks where the cyan is 50% (actually 49.8%) and adjacent color blocks are measured with a spectral photometer to find a neutral gray value that is closest to the target (i.e.0 a*,-1 b*). If the middle color block is closest to the target gray, the equipment is already gray-balanced (at 50% C) without any correction. If the a*b* value closest to the target is not the middle color block, pay attention to the percentage values listed next to M and Y. For example, if the optimal measure is between+2 and+3 of M, which is obtained on -3Y, the preferred gray balance desired is +2.5M and -3Y. By repeating this step, the actual gray balance value can be found for the color blocks such as 75%, 62.5%, 37.5%, 25% and 12.5%, etc.

② Determine the NPDC curves of single-color C, M, and Y: On the CMY plot, draw the curve of the fourth column value of P2P, you will get the C plate curve; then draw the curve of single-color M and Y plate to the left or right of the original curve (C plate) by the percentage found on GrayFinder.

③ Determine the standard NPDC curve and correction points, and draw a new NPDC curve.

2. Determine the standard curve of the printed solid density value

In the NPDC sector graph, determine the standard curve that is closest to the actual printed solid density value. Then, locate the point that needs to be corrected. Calibration is made for each correction point. Draw a vertical line from bottom to top, and intersect the target line at the intersection point. Draw a horizontal line that intersects the C, M and Y lines at the intersection. Draw many vertical lines from top to bottom and intersect with the coordinate horizontal axis to obtain three new target values of C, M and Y. The target value of CMY is recorded on the new value column of the drawing to be repeated for each curve point, 0% and 100% are unchanged.

3. Calibration of Mono-colored Black Edition

On the black special drawing, draw the value of the fifth column of P2P on it. Please refer to NPDC correction for three-color CMY as a practice.

4. Assign a value to RIP

The NPDC correction results of CMY and K above are assigned new target values for the RIP or correction device. Some RIP devices require the users to enter the value "after measurement" rather than the value "required", and others require them to enter the value of the correction difference. The new target value is the value that should be obtained for each curve point after correction.

IV. Printing

1. Platemaking

A new printing plate of the standard proofs is made with a new RIP curve, and the value of the printing plate on P2P is compared with the recorded unadjusted printing plate curve to ensure that the required changes have been obtained. For example, if 50% of the curve points have a new target value of 55%, then the users need to check whether the 50% color blocks of the new plate are about 5% larger than the unadjusted one. It is difficult to measure the surface of printing plate, so the values don't need to be extremely precise.

2. Printing

Use a new printing plate or RIP curve and print the standard sample with the same printing conditions. It is best to be the whole measurement standard sample. Printing according to the same L*a*b* value (or density value) as recorded at the end of the correction printing. Note the uniformity and gray balance of ink color. When the machine is debugged, HR, SC, and HC values are measured to determine that the printing press meets the NPDC curve. Also measure the P2P standard sample, if possible, or hand-draw the 4th and 5th columns on an empty G7 drawing. These curves should be nearly consistent with the target curve at this time. If not, adjust the solid density, or print a few more copies for a printing press to warm up. Check other parameters, such as gray balancing, uniformity, etc., whose data are under control. Then turn the machine on at normal printing speed and check whether the measured value is ideal throughout the process. At least two or more sheets should be picked from the site and dried naturally. If possible, perform printing press operation twice or more times under the same conditions, and pick the best printing sheet from each printing to prepare for subsequent work averaging.

3. Establish ICC

The characteristic data of each printing sheet selected was measured with a spectral photometer; then a ICC file of the printing press was created from the average data. Store the original measured spectral data, rather than the CIEL*a*b* (50), if possible. By obtaining the improved ICC file in this way, the metamerism caused by changes in non-standard light sources or both light sources can be reduced.

Chapter 9 Lithographic Printing Standards and Management

Section 4 Production and Environment Management

I. Related Knowledge

1. Related Standards of Environmental Management System

The ISO 14000 series of environmental management standards are a series of international management standards established by International Organization for Standardization (ISO) on the basis of the successful development of ISO 9000 family standards. Currently, the latest standard of ISO 14000 standard is the 2015 edition, and China has adopted the standard and promulgated GB/T 4001-2016 *Environmental Management System Requirements and Use Guidelines*.

(1) Composition of ISO 14000 Series Standards

ISO14000 series standard is an international standard drafted by ISO/TC 207, the International Organization for Standardization. ISO 14000 is a series of environmental management standards that includes many key issues in international environmental management fields, such as environmental management systems, environmental audits, environmental indicators, and life cycle analysis, and is designed to guide organizations (enterprises, companies) in achieving and performing proper environmental behaviors.

ISO 14001 is the leading standard of the ISO 14000 series standards. Its overall purpose is to support environmental protection and pollution prevention and promote the coordinated development of environmental protection and social economy. In this regard, ISO 14001 standard emphasizes on the requirements of pollution prevention and continuous improvement. It also requires to control environmental factors and reduce environmental impact in all aspects of environmental management, and the idea and method of pollution prevention throughout the establishment, operation and improvement of the environmental management system. Currently, modern enterprise is implementing the ISO 14001 environmental management system standard.

(2) Contribution and Significance of ISO 14000 Series Standards

① They contrubute to improving the environmental awareness and management level of the organization

The ISO 14000 series of standards are systematic standards for environmental management. They are a set of complete and operational system standards formed by integrating the experience of many developed countries in environmental management. During the implementation of the environmental management system, enterprises shall first evaluate their current environmental situation and determine major environmental factors; plan their issues in all aspects and at all levels, such as products, activities, services, etc., and carry out training, operation control, monitoring and improvement through a documented system, and implement process control and effective management. Besides, an environmental management system should be established to enable enterprises to gain further insight into environmental protection and its intrinsic value, and enhance their sense of responsibility for environmental protection in production activities and services. Adequate awareness is formed of the environmental factors that exist and potential in the activities of the enterprises and related parties. As an effective means and method, this standard establishes a systematic management mechanism based on the original management mechanism of enterprises. New management mechanisms will not only improve environmental management level but also promote the overall management level of enterprises.

② It is conducive to the implementation of clean production and pollution protection

The ISO 14000 environmental management system emphasizes pollution prevention and clearly specifies that enterprises must make a commitment to pollution prevention in their environmental guidelines, which promotes the application of clean production technologies; comprehensively identifies environmental factors in their activities, products and services in the identification and evaluation of environmental factors. The possible environmental effects of different states and tenses of the environment, as well as the effects of pollutants emitted to the atmosphere and water bodies, noise, and the treatment of solid waste, should be investigated and analyzed item by item. Existing issues should be solved managerially or technologically to be incorporated into the management of the system. Management of these pollution sources is carried out through control procedures or project instructions, which embodies the principle of pollution prevention by controlling pollution from the source.

③ It is beneficial for enterprises to save energy, reduce emissions, and reduce costs

The ISO 14001 standard requires effective control over the whole production process of enterprises to reflect the idea of clean production. Environmental factors such as reducing the generation, emission and environmental impact of pollutants, saving energy, resources and raw materials, and recycling of waste are considered from the initial design to the final product and service. In addition, key environmental factors are controlled through the setting of targets, indicators, management plans and operation control, so as to effectively reduce pollution, save resources and energy, effectively use raw materials and recycle waste materials, and reduce various environmental costs (investment, operation costs, fines, sewage charges), thus significantly reducing costs and obtaining not only environmental benefits but also significant economic benefits.

④ Reduce pollution emissions and reduce the risk of environmental issues

The ISO 14000 standards emphasize on pollution prevention and full process control, so pollution emissions can be reduced from all aspects through the implementation of the system. Many enterprises avoid the emission of pollutants through the operation of the system, some reduce the emission of pollutants through the improvement of product design, process and management, and some make the emission of pollutants up to the standard through management. In fact, the role of ISO 14000 is not just to reduce pollutant emissions. In a sense, it is more important to reduce the occurrence of liability incidents. Therefore, the establishment and implementation of the system enable each organization to be fully prepared and properly managed for its own potential accidents and emergencies, which can greatly reduce the occurrence of responsible incidents.

Chapter 9 Lithographic Printing Standards and Management

⑤ It ensures the companies be compliance with laws and regulations and avoid environmental criminal liability

Now, various new laws and regulations around the world are constantly introduced and are becoming increasingly strict. An organization can only obtain these requirements promptly and ensure to be compliance with them through the operation of a system. Also, due to proper control and management, major accidents can be avoided and finally environmental criminal responsibility can be avoided, too.

⑥ Satisfy customer requirements and increase market share

Currently, ISO 14000 standard certification has not yet been one of the market access conditions, but such requirements have been made by many enterprises and organizations against suppliers or partners. Some internationally renowned companies encourage their partners, in accordance with ISO 14001, to seek registration of this international standard by comparing their environmental management systems, implying that priority will be given to suppliers officially implementing ISO 14001.

⑦ Obtain a green pass to enter the international trade market

In the long run, the impact of ISO14000 series standards on international trade cannot be underestimated. Now, most of the "green barriers" realized in the international market are environmental protection requirements of products or production processes put forward by enterprises to suppliers. The ISO 14000 series standards will become one of the basic conditions in international trade.

It will be an opportunity for developing countries to break down trade barriers and enhance their competitiveness by implementing the ISO 14000 series standards.

The ISO 14000 series of standards provide an effective environmental management tool for organizations, especially for productive enterprises. Enterprises that implement standards are generally reflected in improvements in management, energy saving and cost reduction, which improve the competitiveness of their products in the international market.

(3) ISO 14001: 2015 Environment Management System Standard

See relevant standards.

2. Platemaking equipment's utilization rate and cost control method

The equipment utilization rate is an indicator of equipment utilization in terms of number, time, production capacity, etc. Improvement of equipment utilization rate is an important goal to improve the economic efficiency of printing enterprises, and measures to improve equipment utilization rate are important elements to achieve this goal.

In general, factors contributing to the low utilization rate of platemaking equipment include outages caused by platemaking failures; adjusted non-production time of platemaking equipment (i.e., preparation time for platemaking); equipment idling and short stops; production of defective products and reduction of equipment production (loss caused by a startup). For overall improvement of the production utilization rate of platemaking equipment, four aspects can be taken to: maximize the efficiency of platemaking equipment, formulate a complete and effective Preventive Maintenance Plan for platemaking equipment, establish an implementation team to continuously improve the utilization rate of platemaking equipment, and implement comprehensive production quality management and incentive policies within enterprises.

The cost of platemaking equipment includes not only the one-time purchase of equipment but also the maintenance, unkeep, repair, etc. during use. The cost control of platemaking equipment is to control the purchase, use, preservation, maintenance and the full play of capacity of platemaking equipment. Management can be strengthened in the following aspects.

(1) The decision to import platemaking equipment should be correct

The first key condition and prerequisite for printing enterprises to control costs is whether the equipment import decision is correct. Printing enterprises shall conduct a comprehensive analysis of the performance, price point, accessories, repair, and service costs of the proposed equipment when purchasing platemaking equipment. Equipment should adapt to the relationship between living sources and supply and demand in the printing market where it is located in order to increase the operating rate of equipment and reduce shutdown costs. It is also about the life cycle of the equipment and is purchased in the dominant age group of the equipment.

(2) Daily operation and maintenance of platemaking equipment should be carried out

The enterprise equipment management department or production department shall formulate the daily management and maintenance system of the platemaking equipment, and supervise the production implementation process of the workshop; there shall be a written system for daily maintenance, such as timing/positioning view, cleaning, etc.; The department also need formulate the specific scope and contents of weekly, monthly and annual maintenance; They need grasp the maintenance time flexibly, which can be appropriately extended during a busy time, carefully overhauled during idle time, and must be fully prepared from the equipment level.

(3) A three-level standard catalog should be formed for the purchase and preparation of vulnerable components

Parts and perishable parts should be purchased in advance according to different time limits and supplemented at any time according to the minimum inventory to ensure continuous production. A pre-plan is required for the disposal of non-vulnerable parts in case of emergency damage, with the shortest supply path and time; it is best to have more than one supply channel to achieve double insurance.

(4) The design capability of the equipment should be maximized

To adapt to the fierce competition in the market, printing enterprises have generally reduced the discount period of printing equipment. Therefore, it requires us to give full play to the production capacity of platemaking equipment in a limited time. The use efficiency of platemaking equipment should be improved as much as possible based on product conditions and licensing conditions after equipment debugging and break-in.

3. Management knowledge of energy conservation and emission reduction

In a broad sense, energy conservation and emission reduction refer to the conservation of material and energy resources to reduce the emission of waste and environmentally harmful substances (including three wastes and noise, etc.); in a narrow sense, energy conservation

Chapter 9 Lithographic Printing Standards and Management

and emission reduction refer to energy conservation and reduction of environmental harmful substance emissions.

The *Energy Conservation Law of the People's Republic of China*, enacted as early as 1997 and officially implemented in 1998, gives energy conservation its legal status. It involves energy conservation management, rational use of energy, promotion of energy conservation technology progress, legal liabilities, etc. The Law clarifies the guidelines and important principles for the development of energy conservation in China and establishes a series of legal systems such as reasonable energy use evaluation, energy conservation product labels, energy conservation standards and energy consumption limits, elimination of backward and high energy consumption products, energy conservation supervision and inspection, etc.

Printing enterprises, as service processing enterprises, are not the main energy consumers in industrial enterprises. Therefore, the key to energy conservation of printing enterprises is to reduce losses and waste in all aspects of energy use in production and to improve their effective utilization. Energy-saving measures for printing enterprises are as follows:

(1) Thermal Energy Conservation

The main forms of heat energy are heating for a certain part of the printing process, heat treatment of raw materials and products, and winter heating of corporate structures. The key to energy efficiency is to improve the efficiency of the heat exchange process and to use as much low-crystalline heat as possible, especially waste heat. For example, printing enterprises' prepress platemaking development, offset printing ink drying, concave printing color dryness, wireless glue bonding and other links are all links that can save heat and make use of waste heat.

(2) Power conservation

As the most widely used secondary energy source in the world, electric energy has been widely used. However, there is an inevitable loss in transmission and use. Improvement of power transmission efficiency and utilization rate of electrical equipment will play an important role in energy saving. Printing enterprises, as end consumers, should take measures to eliminate low-efficiency motors or high-powered equipment, transform the original motor system adjustment mode, promote advanced power consumption technologies such as frequency conversion speed regulation and independent drive, and choose a reasonable electric heating mode to reduce electric heating loss.

(3) Hydropower conservation

The shortage of water resources is a serious constraint on the economic and social development of China, especially in the north. China has entered a period of water shortage. Printing enterprises use a large amount of water for film development, printing plate development, printing press circulating cooling, printing workshop cooling, etc. in production. It is necessary to improve the energy utilization efficiency of water-used equipment and adopt new processes to reduce the effective water produced by-products so that water energy can be directly saved.

(4) Reduce energy consumption

Scientific organization and management of enterprises should be strengthened to reduce the consumption of raw materials through various approaches, such as paper, ink, wetting liquid, plate materials, film, rubber rollers, blankets, car washing water, glue and film. It is necessary to reduce both direct and indirect energy consumption in printing while ensuring the quality of printing.

(5) Increase the utilization rate of printing equipment

The investment in advanced printing equipment and the annual cost of equipment preservation and maintenance is significant. How to make full use of printing equipment, give full play to all functions of printing equipment to maximize its use, and reduce non-working hours of printing equipment, especially costly maintenance are all factors related to the full use of printing equipment.

II. Steps

(1) Get insights into the processing process and product quality requirements.

(2) Analyze and control factors of total quality management.

(3) Identify elements to be adjusted or focused on control during implementation to stabilize product quality.

(4) Implement total quality control and integrated management.

III. Precautions

(1) All practitioners should actively study *the 13th Five-Year Plan for the Development of National Environmental Protection Standards* issued by the Ministry of Ecology and Environment of the People's Republic of China.

(2) All units should carry out various forms of publicity and education activities to popularize knowledge of green printing and enhance the green printing consciousness of practitioners in the industry.

References

[1] Mu Jian. *Applied Computer Technology of Prepress*[M]. Beijing: Posts & Telecom Press, 2008.

[2] Helmut Kipphan. *Handbook of Print Media*[M]. Xie Punan, Wang Qiang, Trans. Guangdong: World Publishing Corporation, 2004.

[3] Tian Quanhui, Zhang Jianqing, Mo Chunjin. *Color Control Technology of Printing*[M]. Beijing: Printing Industry Press, 2014.

[4] Zhao Guang, Yao Leilei. *ISO 12647-7 New Standard for Digital Proofing*[EB/OL]. (2017-10) [2023-10]. http://www.keyin.cn /people /mingjiazhuanlan /201710/30-1107622.shtml.

[5] Shen Weizhi. *Color Management Certification for Digital Proofing*[J]. *Digital Printing*, 2010.